AF740498

MEN AND MACHINES

Hywel Murrell

DE MONTFORT UNIVERSITY
LIBRARY:
B/code
Fund: Date: 8/8/06 3 19
Sequence:
Class: 150
Suffix: MUR

Methuen

To Ruth

First published 1976 by Methuen & Co Ltd
11 New Fetter Lane, London EC4P 4EE
© 1976 Hywel Murrell
Printed in Great Britain by
Richard Clay (The Chaucer Press), Ltd
Bungay, Suffolk

ISBN (hardback) 0 416 82300 9
ISBN (paperback) 0 416 82310 6

We are grateful to Grant McIntyre of
Open Books Publishing Ltd
for assistance in the preparation of this series

This edition is available in both hardback and paperback editions. The paperback edition is sold subject to the condition that it shall not, by way of trade or otherwise, be lent, re-sold, hired out or otherwise circulated without the publisher's prior consent in any form of binding or cover other than that in which it is published and without a similar condition including this condition being imposed on the subsequent purchaser.

Contents

Acknowledgements

I would like to thank Mrs Elizabeth Clarke who transcribed the whole of this book from tape and the other clerical staff of the Department of Applied Psychology at UWIST for assistance in retyping. I particularly wish to thank my wife and youngest daughter for carrying the burden of this spring's lambing so as to leave me free for writing.

The publishers and I would also like to thank the following for permission to reproduce some of the figures in the text (for full bibliographic details please see References and Name Index):

The Editor of the International Journal of Production Research for Figs. 6.2, 6.3, 6.4, 6.5, 6.6, and 6.7; Chapman and Hall for Figs. 2.2, 2.3, 2.4, 2.5, 2.6, 2.7, 3.1, 3.2, 3.4, 3.9, 4.2, 4.3, 4.6, and 6.1 from *Ergonomics* (1965); McGraw Hill for Fig. 3.3; and for Fig. 3.8. the British Standards Institution, 2 Park Street, London W1A 2BS, from whom complete copies can be obtained,

DMU 0554842 01 4

ESSENTIAL PSYCHOLOGY

General Editor
Peter Herriot

E4

MEN AND MACHINES

ESSENTIAL

A

1 An Introduction to Psychological Science/*David Legge*
2 Physiological Psychology/*John Blundell*
3 Learning and Reinforcement/*Stephen Walker*
4 Perception and Information/*Paul J. Barber and David Legge*
5 Information and Skill/*David Legge and Paul J. Barber*
6 Human Memory/*Vernon Gregg*
7 Thinking and Language/*Judith Greene*
8 Experimental Design and Statistics/*Steve Miller*
9 First Experiments in Psychology/*John Gardiner and Zofia Kaminska*

B

1 Social Behaviour/*Kevin Wheldall*
2 Interpersonal and Group Behaviour/*Judy Gahagan*
3 Values, Attitudes and Behaviour Change/*Ben Reich and Christine Adcock*
4 The Social Psychology of Organizations/*Frederick Glen*
5 Psychology and Social Structure/*Barrie Stacey*

C

1 Growing and Changing/*Harry McGurk*
2 Cognitive Development/*Johanna Turner*
3 Growing Up in Society/*Tony Booth*
4 Culture's Influence on Behaviour/*Robert Serpell*
5 In and Out of School/*Joan Freeman*

D

1 The Person in Psychology/*John Radford and Richard Kirby*
2 Motivation/*Phil Evans*
3 Approaches to Personality Theory/*David Peck and David Whitlaw*
4 Individual Differences/*Richard Kirby and John Radford*

E

1 Psychology and Work/*D. R. Davies and V. J. Shackleton*
2 Selection and Assessment at Work/*Gilbert and Helen Jessup*
3 The Psychology of Training/*Robert Stammers and John Patrick*
4 Men and Machines/*Hywel Murrell*
5 Motivation at Work/*Hywel Murrell*

F

1 Need to Change?/*Fay Fransella*
2 The Psychology of Handicap/*Rosemary Shakespeare*
3 Clinical Psychology: Theory and Therapy/*Dougal MacKay*
4 Community Psychology/*M. P. Bender*
5 Psychology and the Environment/*Terence Lee*
6 Consumer Psychology/*Mary Tuck*
7 Images of Man/*John Shotter*
8 Radical Perspectives in Psychology/*Nick Heather*

PSYCHOLOGY

Editor's Introduction

Some typewriters have been designed, as Hywel Murrell points out, to be operated by ambidextrous praying mantises. This thorough book indicates the extent to which we have applied knowledge about human skills to work situations. We have to consider particular characteristics of men and machines, and also how a man–machine system works as a whole. It is another case of human beings changing their environment and responding to the effects of that change in their subsequent behaviour.

Unit E of *Essential Psychology* deals with a particular area of applied psychology: pyschology and work. This is an ideal topic to demonstrate the ways in which the different psychological models of man and their associated findings can be utilized. The human being as processor of information copes with varied inputs in his job and has to adjust his behaviour accordingly. As a social being, he is much affected by the groups of people with whom he works and the organization within which he operates. As a developing person, his changing view of himself will partly consist of his view of himself at work. As an individual, he brings differing skills and motives to his work. Above all, as a human being, he possesses the capacity to change his work situation to suit his own abilities and objectives.

Essential Psychology as a whole is designed to reflect this changing structure and function of psychology. The authors are both academics and professionals, and their aim has been to introduce the most important concepts in their areas to beginning students. They have tried to do so clearly, but have not attempted to conceal the fact that concepts that now appear central to their work may soon be peripheral. In other words, they have presented psychology as a developing set of views of man, not as a body of received truth. Readers are not intended to study the whole series in order to 'master the basics'. Rather, since different people may wish to use different theoretical frameworks for their own purposes, the series has been designed so that each title stands on its own. But it is possible that if the reader has read no psychology before, he will enjoy individual books more if he has read the introduction (A1, B1 etc.) to the units to which they belong. Readers of the units concerned with applications of psychology (E, F) may benefit from reading all the introductions.

A word about references in the text to the work of other writers – e.g. 'Smith (1974)'. These occur where the author feels he must acknowledge an important concept or some crucial evidence by name. The book or article referred to will be listed in the References (which doubles as Name Index) at the back of the book. The reader is invited to consult these sources if he wishes to explore topics further.

We hope you enjoy psychology.

Peter Herriot

I
Introduction

The student of psychology, and other readers of this series, may perhaps be excused for wondering why a book with the title 'Men and Machines' should be part of a series on Essential Psychology. After all, the psychologist (traditionally at any rate) is supposed to be concerned with the functioning of the human mind and the behaviour which follows from it, whereas machines are the province of the engineer. But machines cannot get on without men. The most expensive Rolls Royce would sit in its garage until it rotted away if no one got in to drive it, and even the most sophisticated computer would produce nothing of any value if someone had not programmed it, if someone did not maintain it and if someone did not look after the machinery which provides the electricity on which it is to run. By the same token man cannot get on without his machines. Take our machines away from us, and our standard of living falls to a level which none of us could tolerate. So man and his machines are very intimately connected; the activities of both must be very closely integrated if optimum results are to be obtained. Expertise on machine design must clearly come from the engineer, but equally clearly, the expert on the part played by man will be the psychologist.

Recognition of this marriage between engineering and psychology is only very recent, and even now is not as extensive as it should be. This may be partly because engineers, since they are men themselves, feel that they know all about men and do not require advice from others and because psychologists are unwilling to come down from their ivory towers, and to get their hands dirty with such mundane things as work on the shop floor or the machines to be found there. Fortunately, the situation is slowly changing, though the drive has largely come from the development of ergonomics in which, of course, psychologists have played an important part.

Until comparatively recently, machines were constructed with little or no regard for the individuals who were expected to work on them. Furthermore, machines were usually the limiting factor in performance rather than the operatives and any limitations which the operatives might have had were largely physical. In recent years, however, the situation has changed quite rapidly; the machines have become increasingly sophisticated and the people working on them have acted less and less as doers and more and more as controllers. And this in turn has shifted the emphasis from the physical aspect of human work to the mental aspect – which clearly will involve psychology. It should be emphasized at this point that the study of men and machines cannot and should not be carried out in isolation from the remainder of the working situation in which this man–machine combination forms a part. So the psychologist, in addition to having association with the engineer, must also have contact with work physiology, with work study and with management – both personnel and general. Some of these aspects are dealt with in other books in Unit E and will not be further expanded here.

The history of the development of psychological interest in industrial activity, including machine design, is very well documented for the United Kingdom and shows a continuity which appears to be less well documented for other countries. This is not to say that interest has not developed elsewhere, but its development seems to have been rather more recent and, so far as continental Europe is concerned, interest in human work stems more from the medical profession than from psychology.

From the turn of the century until the outbreak of World War One, there was developing a new approach to psychology. Members of the great school of German psychologists were becoming old and the new concept of experimental psychology was straining to break away from the confines of philosophical psychology to give a fresh approach to the solution of psychological problems. The idea that experimental results could be applied to real life situations very quickly developed and the word 'applied' began to appear, sometimes in conjunction with 'experimental' and sometimes on its own. As early as 1913 Münsterberg published a book entitled *Psychology and Industrial Efficiency*, in which he suggested that various psychological techniques could be used to achieve various industrial ends – particularly the selection of the best possible individuals for jobs and the establishment of the best possible working conditions. Just prior to this, in 1912, C. S. Meyers had set up the Experimental Psychology Laboratory at Cambridge which some ten years or so later was to be taken over by Sir Frederick Bartlett, the doyen of experimental psychologists who did so much to inspire work on the relationship between man and his equipment.

After the conclusion of World War One, two separate but closely linked developments took place. The Industrial Fatigue Research Board (IFRB) was set up and the National Institute of Industrial Psychology was established by Meyers. The former was fragmented during World War Two, one offshoot becoming the Applied Psychology Unit, located at Cambridge under Bartlett; the latter ceased trading in 1973. In the early days both were concerned to some extent with machine design and the problems of man–machine interaction. But, as time passed, they showed much greater concern for the personnel side of occupational psychology and it was not until the war produced a requirement for studies of man–machine interaction that this side of the work was developed to any great extent and it is only in the APU that it has persisted.

But even so, it was not on the British side of the Atlantic that the greatest developments took place. Both the US Army Air Force (Fitts, 1947) and the Office of Naval Research (Taylor, 1947) lavishly supported psychological research into equipment design, both in their own establishments and in universities such as Johns Hopkins, Tufts and Princeton. The Americans quickly gave the name Human Engineering to this area of endeavour but this was subsequently changed to Human Factors Engineering, and more recently to Human Factors. The British forces were rather less lavish in their support, but they did set up Personnel Research Committees, part of whose activities were concerned with equipment design; the Navy had its Operational Efficiency Sub-Committee which had close liaison with the APU at Cambridge while the Army had a section of the Army Operational Research Group. The Air Force developed its Institute of Aviation Medicine at Farnborough which, despite its name, had a number of psychologists on its staff.

With the development since 1949 of ergonomics, with its interdisciplinary approach, human factors in equipment design has ceased to be the almost exclusive province of psychologists. Furthermore, less work is being done by government-supported bodies and more is being pursued by research in universities such as the Department of Applied Psychology in Aston, or by private enterprise such as at Pilkingtons, St Helens or EMI at Feltham. Although people undertaking this research may well have come from other disciplines, the techniques which they are using will be largely derived from the methods developed by applied psychologists.

Engineering designers were largely concerned with producing equipment which was either convenient to manufacture or conformed to previous practice, irrespective of whether a person who actually operated the equipment could do so efficiently. Therefore the early psychological work was largely concerned with minutiae of design parameters, e.g. the design of indicating instruments like dials, the layout of panels, the design characteristics of control knobs and so on. More recently, however, much interest has developed in man's role in complex systems. This has involved studying what man can and what he cannot do well; what is best done by a machine and what is best done by a man; and, extending beyond this, how long should he do it before his efficiency falls off. Another more recent line of enquiry has been into the problems posed by production lines, particularly where work is machine paced. It is true that many of these latter areas were touched upon by the IHRB workers between the wars. But the main difference between their work and that which is going on now is that theirs consisted primarily of shop floor observation; recent research is much more concerned with laboratory studies, which are attempting to tease out fundamental principles which may underlie field observations. Another post-war development has been interest in training as it relates to equipment design and vice versa (see E3).

This then is a very brief outline of psychology's involvement with the design of mechanical hardware. Perhaps because more funds were available or because they think the prize is greater, the Armed Forces seem to have been much more concerned than has been private industry. Industry has been much more desirous of making people work on the equipment which exists, rather than matching the machines to the capacities of men. It is true that there have been many developments in industry, from the early days of Taylor in 1887 with his scientific management, through Gilbreth with his motion study in the Edwardian period to Herzberg in the 60s (see E1), but on the whole, these efforts have been less directed towards achieving optimum man–machine relationships than in producing high 'motivation' to produce more. This latter aspect is dealt with in greater detail in E5.

Clearly the research effort, mainly over the last twenty-five years, has produced a great deal of relevant data. But it is not the intention that this book should be a kind of 'cook book' setting out a plethora of design recommendations for study by student or designer; this has already been done competently and fully elsewhere (e.g. Murrell, 1965a or Van Cott and Kinkade, 1972). Instead various lines of research in particular areas, and usually by different investigators, will be reviewed with comments on the methods used, the results obtained and where appropriate the practical outcome. The object is to concentrate more on the psychological approach to investigation and realization of the relationship between men and machines than upon the design detail which may result from the research.

2
Man controlling machine

Before going into detailed consideration of the relationship between men and machines, it is necessary to understand the relative parts played by the man himself and by his machine. In any activity a man will receive and process information and then act upon it. Similarly, the machine will receive instructions, carry out its allotted task and will in one way or another inform the man of the progress of what it is doing. In this context the word 'machine' is used to imply any piece of equipment with which an individual may accomplish some purpose. The pencil with which we write, the racket with which we play tennis or the fork with which we dig the garden are in this sense just as much 'machines' as the car we drive or the lathe on which we may be working.

When operating a machine or, perhaps better, when operating *with* a machine, man will receive information through his receptor function, that is, largely through the sense organs of the eyes and the ears, but also through the sense of smell, through touch, through sensations of heat or cold or through kinaesthesia (see Fig. 2.1). The incoming information will be transmitted through the nervous system to the central mechanism of the brain and spinal chord, where it will be processed to arrive at a decision. The processing may involve the integration of incoming information with information which has already been stored in the memory and the decisions which will result may vary from responses which are virtually automatic to those which involve a high degree of reasoning or logic. Having received the information, processed it and arrived at a decision, a man will then take action; he will do this through his effector mechanism which involves some form of neuro-muscular activity based on the skeletal framework of the body. The machine

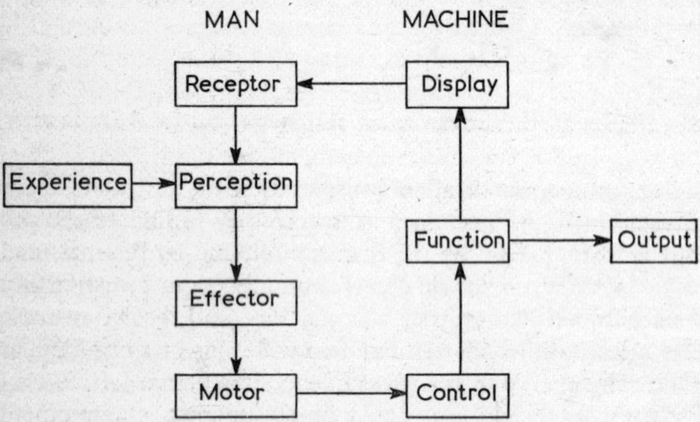

Fig. 2.1 *Man as controller in a closed loop man–machine system*

will receive these commands through its control mechanisms. It will then carry out any activity which is commanded of it to produce an output and will inform the operator of what is going on by means of some form of display. To achieve maximal efficiency, a man–machine unit must be designed as a whole with the man being complementary to the machine and the machine being complementary to the abilities of the man.

A man–machine unit can function in two ways: the first, which is becoming the most common in manufacturing industry, is where the man gives the machine a command to start and he then sits back and leaves the machine to run itself. It is only if something goes radically wrong or when the task is complete that the machine will have to communicate with the man: 'hi, I am in trouble' or 'I have finished what you have told me to do'. If the man is hanging around waiting for a communication from the machine in this sense, he becomes a monitor or watch keeper or in more common parlance a machine minder. The second way in which a man and machine may interact is where the man actually 'drives' the machine, as in the transport industry. He gives the machine a command and as a result the machine takes some action and immediately and continuously informs the man of what it is doing. As a consequence the man may give further commands to the machine which may result in a change in the information which the machine conveys to the man. In this function, the man and his machine are acting as a closed-loop servo-system displaying many of the feedback characteristics of such a system. Moreover, the man will usually form that part of the system which makes decisions and so the efficiency of the system will usually depend very largely upon how well the man carries out his 'black box' functions. The speed with which a system functions may be quite high, as when driving a motor car or aeroplane, slower, as when steering a ship, or quite slow as when a process worker is operating a chemical plant which is under manual control.

Not all man–machine units will act in this way. The relative roles of the man and the machine may be reversed. It may be the machine which gives the command to the man to take some action; it may give the man a limited time in which to take this action and if the man does not comply it may go inexorably on extracting some kind of penalty. Arrangements of this kind are usually called paced and are to be found on many assembly lines where the penalty for failing to keep up with the machine is either that the operative gets out of station or units of production are lost. There are situations in which the man can stop the machine until he has caught up, the penalty in this case also being a loss of production. A cigar-wrapping machine is an example of this; an arm swings over at about 15 times per min to pick up a wrapper leaf which has been spread over a die. If the spreading process is not completed in time the operative can stop the swinging arm with a foot pedal but she has to be pretty smart about it or she could get her knuckles rapped. Finally there are situations in which the machine will simply call for attention and may exact no penalty if it is not satisfied. Such a situation would be found on a private branch telephone exchange where the operator's attention is called either by a light or a buzzer to the effect that someone wants to communicate with her. If she decides not to answer, but rather to continue powdering her nose or talking to her boyfriend, the machine itself will exact no penalty. What her employer does is of course another matter.

To evaluate the role of man in all these different situations, it is necessary to know something about the nervous system, about the functioning and capacity of the central mechanism and, invading the realm of functional anatomy and work physiology, the structure of the body – of the bones and the joints and of the muscles which provide the motive power which is man's sole output. In addition, something needs to be known of the source of power which drives this mechanism, and the limits of output which can be expected from it. The man–machine does not exist in a vacuum; any of the psychological or physiological functions of the man may be influenced by working in an environment which is too cold, too hot or just right; he may be subjected to noise which may be of such intensity and duration that he may suffer physical impairment of hearing or his means of communication may be seriously interfered with. To see, he must have light which should be of a quality and quantity adequate for the needs of the job. His work must be organized so that he can maintain maximum efficiency and interest and so that his abilities are being utilized to the full. His relations with other members of his working group should be such that his efficiency is not impaired.

In all these different aspects of man and his machine, there is immense scope for investigation by the applied psychologist. In general, this investigation may take two forms: first, general principles which can be applied in a wide variety of situations may be examined in order to produce design recommendations of wide applicability. Secondly, specific pieces of equipment, or families of pieces of equipment, may be examined in great detail and the outcome will probably not have any application beyond the kind of machine under investigation.

Allocating functions between man and machine

When equipment is intended for human use, it should, as far as possible, be designed as a man–machine unit. There are times, however, when this may involve a compromise between the needs of the man and the requirements of the machine. This compromise may in turn require the striking of an economic and operational balance between the two.

In the first instance, it will be necessary to decide whether a particular function can best be carried out by a man or by a machine. In making this decision, the functions which man can perform well will be weighed against the prime cost of replacing him by a machine as well as against any supposed operational advantage. There are some things which a man can do very well, better than most machines; conversely there are other functions at which he is rather poor and which should be given to the machine wherever possible. Once the relative roles of the man and the machine have been decided upon, the equipment must be designed so that the man is given the greatest opportunity to function efficiently at all times. He must work in the best conditions and his activity must be organized in due regard of the demands made upon him by the job.

The next step in planning a man–machine unit will usually be to decide what information the operator will need in order to carry out his task and how best he can get this information; he may be able to see things happen or he may need some form of visual, auditory or tactual display. If some form of display is required, it should be chosen to give the information as quickly and as unambiguously as possible. The limits of accuracy in assimilating the displays must be decided upon and the types of display which can give this information most accurately under particular circumstances must be chosen. The need of communication between other operators in similar units must not be overlooked. Some form of control will almost certainly be needed and the degree of speed and accuracy likely to be demanded in their operation must be assessed. The physical loads which are likely to be involved and the possible need for several assistants when the load is heavy will have to be determined. The kinds of control which meet the requirements can then be chosen and as far as possible both they and the display must be designed around the operator and in relation to each other. Some appraisal of the physical conditions which are probable will have to be made, especially if the equipment itself is likely to contaminate the ambient environment with heat or noise.

Early consideration will have to be given to the general mental demands of a task of actually operating equipment so as not to overload the operator. An overload situation can develop unless information reaches him at a speed with which he can cope and in a form which he can easily assimilate. It should be properly spaced in time and properly located around him so that he can act upon it without undue mental stress. As far as possible, he should not be forced to work at a fixed pace.

Although there is some psychological research to support the foregoing assertions, on the whole most of what has been said is either the result of experience (applied 'common sense', if you will) or is the result of careful extrapolation from research carried out for other purposes. There are great practical difficulties in investigating these matters. It is frequently possible to find situations in which violation of the ideas which have been put forward can be demonstrated, but it is not always easy or practical to persuade an enterprise to redesign jobs so that you can get a comparison. Not unnaturally a redesign will often include technological improvement at the same time (and you can't blame a firm for doing this) so that a comparison of like with like is virtually impossible. In the same way it is not possible to construct sub-optimal man–machine units simply so that they can be compared with a situation which is better designed; the cost would be prohibitive. So, much of what has been asserted will have to be accepted as an act of faith without specific laboratory evidence for most of the points which will be made. Much of the evidence has been produced, a good deal of it indirectly, from work on man as a controller of complex systems going back to the Second World War, and in particular the development of the concept of the human controller in non-psychological terms, by such people as Craik (1947a and b), Hick and Bates (1950), Tustin (1947), North (1956) and others in Britain, and in America by workers with Frank Taylor at the Naval Research Laboratory in Washington, much of which was published posthumously after Taylor's untimely death in 1957 (e.g. Taylor, 1959 and Taylor & Garvey, 1959).

Although it will only rarely be done in this way, at the outset of any design process an analysis should have been made of the overall functions of the man–machine unit under consideration and decisions should have been taken on which of those functions will be allocated to the man and which to the machine. This will be part of a process of system design which, in the more complex situation, will demand consideration of performance requirements, cost, reliability and feasibility. The description below is an outline of what can be a complex iterative process. But this will not always be so; in many instances it will be quite obvious who does what. But when it is not obvious analysis will be required as to what things are best done by man and what by machine. Fitts (1951) was the first to draw up a clear list of these functions. Lists derived from Fitts' ideas have appeared in a number of publications, but they all say essentially the same thing with slightly different wording and different presentation. So, following Murrell (1965a), we can set them out below:

Functions which can usually be done better by a man than by a machine
Sensing minimum stimuli
Amplification
Improvisation and flexibility
Switching from one input to another
Long-term storage (memory) of great capacity
Perception of space, depth and pattern
Interpolation
Extrapolation and prediction
Translation
Inductive reasoning
Making complex judgements or decisions
Homeostasis

Functions which it is usually best to allocate to a machine
Computing
Differentiation
Integration
Response at great speed
Application of massive force, smoothly and precisely
Precise repetition
Short-term information storage (memory)
Deductive reasoning
Complex simultaneous functions
Simple yes/no decisions
A man in a system will introduce delays, noise and limits on data transmission.

We can now look at some of these functions in detail.

Flexibility
Flexibility is probably one of the most useful human characteristics. Unlike a machine, which is designed for a specific purpose and which, usually, can work only within pre-programmed limits, a man can change his role rapidly and frequently; he has

a multiplicity of channels which he can select as required and his homeostatic processes enable him to do this effectively under a variety of conditions, favourable and inimicable; he can continue to function even when some channels have been rendered inoperative either temporarily or permanently. Of equal importance is man's function as a sensing device to minimal stimuli from a wide range of sources simultaneously, bringing these signals together to produce a complete picture of an event. In carrying out this sensing he may respond to very dim lights or quiet sounds in an all-or-none function, and he may be able to sense changes in these functions which may be very small indeed and can relate these to changes in other functions so as to arrive at a 'view' of their meaning. Since combinations of changes may be almost infinite in variety, it may be very difficult if not impossible to programme a machine to cover the wide range of eventualities which could be covered by a man. The complexity of some cues used by experienced people is often not realized until attempts are made to determine exactly how a man arrives at his conclusions and on the basis of what information. When the man acts as a result of sensing minimal stimuli he is acting as an amplifier of very high gain to set underway a complex train of consequences. For example, a pianist will amplify the signs on a sheet of music into a complex series of physical movements which can produce for an audience a sense of aesthetic pleasure which, so far, is not attainable by purely mechanical means. It is true that a player piano using a punch tape can play the notes in the right sequence but it is not difficult to distinguish between a work played by a live pianist and one played on a player piano. Alternatively, an operator may be aware of a slight change in colour or in sound which will tell him that something untoward may be happening in the equipment over which he is operating. He will therefore in the light of his previous experience interpret what these changes mean and take the appropriate action.

Other jobs which men do very well and for which no viable substitute has been developed include the dyeing of wool in the manufacture of carpets, where the operative compares the hue of a skein dipped in his vat with that of a sample; he can then judge how much of a variety of dye stuffs must be added to the vat in order to get a colour match. Or the job of cheese grading, which was so extensively studied by Harper (1952), who could find no substitute for the pressure of a man's thumb and the skill of his nose. Even in automatic machines such as letter or parcel sorting it still requires a man to read the post code or the address and to instruct the machine into which channel the missive is to be directed. There are other areas where man's sensing function has been replaced by machines. For instance, the visual lookout has been replaced by radar scanning which will display the echo of an oncoming ship on a collision course, but this echo has still to be detected by an operator and the correct conclusions have to be drawn.

Prediction and storage

Man is good at interpolation and extrapolation, that is, he can predict within the present situation and beyond it. It is true that when the situation is very static this can equally well be done by a machine, but when a situation is very fluid man's greater flexibility may be an advantage. When the predictions involve rate of change so that differentiation or integration is required, the task if at all possible should be given to the machine since these are the functions which men do really rather badly. Nevertheless, there are many situations in everyday life where integration or differentiation is required as, for instance, when driving a motor vehicle. Judgements continually have to be made of relative speeds of overtaking and oncoming motor vehicles: no substitute for man has yet been devised in this context.

Memory may be either permanent or transitory. In the permanent sense data are stored over a long period of time which accumulates to become 'experience'. In the transitory memory are stored data that are required for immediate use and which, when once used, can be discarded. In general, transitory memory is best assigned to a machine; any attempt to replace the whole of man's permanent memory is likely to prove immensely expensive and rather ineffective except for numerical data. Nevertheless, machines should have the maximum viable amount of memory built into them so that they rely as little as possible on the memory of the operator; their operation will then be liable to be learned more quickly and they will be operated more efficiently, particularly if the operator is becoming old.

Decision-making

Finally, we come to man's decision-making function. There are certain semantic difficulties over the word 'decision' since it has a very wide range of meaning, though this does not necessarily mean that the processes involved are very different. When a decision is made by an executive he will weigh up a number of factors and make a prediction of the eventual outcome of the decision. This is very easy to recognize. It is much less easy to recognize the same process in the automatic decisions which are made from moment to moment when a motor response or a series of motor responses are made as a result of incoming information. The exact boundary between decision-making and reflex action is hazy, but many activities which might appear to be automatic do, in fact, involve the making of decisions. These decisions do not differ very greatly in their procedure from the process of decision-making by the executive. It is very largely a question of the time interval involved. Thus the operator will receive data from the environment or the equipment which must be compared with data stored in the immediate or permanent memory; from these he will arrive at a number of alternative outcomes from which he must select one most likely to give the optimum result. In selecting the optimum outcome the operator may be assisted by previous experience which has enabled him to set up a series of rules to assist in reducing the uncertainty of the outcome and increase the speed of the decision. The simpler this set of rules, the quicker can the decision be established; the smaller the number of alternative outcomes, the more efficient is the process of decision-making likely to be.

In carrying out his decision-making process a man is probably performing the function for which he is best suited. But there are very definite limitations to decision-making and these limitations must be taken into account if he is to do his task efficiently.

The single channel function. There is evidence that at the highest decision-making level in the cerebral cortex man acts as a single channel mechanism (Welford, 1959) and that the rate at which data can be transmitted through the channel is strictly limited. When one piece of data enters the channel a period, which has been described as the psychological refractory period (Welford, 1952), inhibits the receipt of further data and the taking of appropriate decisions. It appears that the refractory period is about 500 msec, so that about two 'chunks' of data can be passed through the channel per second. If two 'chunks' of data arrive almost simultaneously, one of them will be passed through the channel, but the other may be delayed, or lost altogether. Data which may be required to be processed simultaneously may be coded in such a way that more than one item is comprised in each 'chunk', but as a general principle, under no circumstances should equipment be designed or a machine cycle evolve which requires an operator to make two simultaneous decisions or two decisions within half a second of each other. This limited rate of transmission has been described by Birmingham and Taylor (1954) as the 'band width' of the human operator.

Of the minimum time of about 500 msec between successive responses only part may be occupied by the central processes. This part will be used by the operator to make a decision and to organize the response, which, once the network has been established, will usually be carried out without further voluntary control on the part of the operator. In fact once the response has been initiated it is virtually impossible for the operator to stop it. The effect of this reaction time, or perhaps more accurately response initiation time, is to introduce delays into a system which must be allowed for in design.

Since the human operator functions as a single channel mechanism with delay, his inputs will be intermittent sampling and he will behave in an intermittent fashion in his responses. Birmingham and Taylor (1954) have described this situation: 'It would seem that if any type of servo-motor is to be taken as an analogue of human behaviour, it would have to be an intermittently sampling servo instead of a continuous follower', and it is in this activity of intermittent sampling that we have an explanation of commonly observed behaviour which might seem to be contrary to the limitations just described. It would explain why it is seemingly possible to read a book and listen to music at one and the same time, that is by alternatively sampling one or the other; if the book is a who-done-it and the music is light and reasonably familiar, alternative sampling is sufficient for the meaning of both to be understood. But if the book be learned and the music be abstruse one can be absorbed only at the expense of the other.

What it amounts to, then, is that if the incoming load be light so that continuous sampling is not required, then there will be what has become known as 'spare mental capacity' available for dealing with other incoming material. For the most part research workers on spare mental capacity have been interested primarily in using it as a measure of the demands of a primary task. The technique was initiated by Bornemann (1942) who studied the interaction of mental arithmetic as a subsidiary task with a number of primary tasks. A systematic laboratory study has been reported by Schouten *et al.* (1962), who used as a primary task the pressing of pedals in response to high or low tones presented in an unpaced or paced condition at different rates representing various percentages of maximum performance. Secondary tasks included putting washers and nuts on screws, solving arithmetical problems and writing. Results showed that simple tasks are relatively unaffected by raising the demand made by the primary task, while complex tasks may disintegrate completely.

The technique has been further developed by Kalsbeek (1964), who has applied it to the establishment of acceptable mental load in air traffic control (Kalsbeek, 1971). Other applications of a practical nature have been reported by Broadbent (1956) on the relative difficulty of listening tasks, Garvey and Taylor (1959) on the comparison of tracking tasks, Poulton (1958), who compared two arrays of dials in order to assess their relative readability, and Poulton and Brown (1961) on car driving.

If a man has to spread his sampling over a wide range of data sources, the amount of data he will receive from any one source will be reduced in proportion to the sampling pattern. This means that the more functions he has to perform, the less data he will have available for the performance of any one of them, and this can introduce imprecision into his performance. But even when he has to sample only one source of information, a man is likely to be less precise than a specialist mechanical or electronic device; the decision as to whether a man should be used or a machine be designed to replace the man therefore depends to some extent on the precision of response required. Furthermore, even when carrying out the simplest functions a man is variable in his performance and this is probably one of his most serious limitations. He will not only vary from moment to moment, but from one period of the day to the next and from day to day (Murrell, 1971). This variability is one of the major causes of the introduction of 'noise' into a closed loop system by the human component. Some of these delays and imprecisions may become less as the operator builds up expectancies based on previous experience in operating the system. Frequent occurrences may have statistical properties which the operator can detect and relationships between sequential events may be developed which can be learned; these together can reduce the randomness of the situation with which the operator is presented and so lead to more effective performance. The building up of expectancies is part of the process of becoming a competent operator and the more the design of equipment helps in this, the greater will be the rate of performance improvement.

Man in a closed loop system

When a man is acting as part of a closed loop unit, the effect of his output may be to produce an immediate change which can be presented to him on a suitable display. In other instances, notably those associated with the control of ships or chemical processes, the dynamics of the process may include a complex

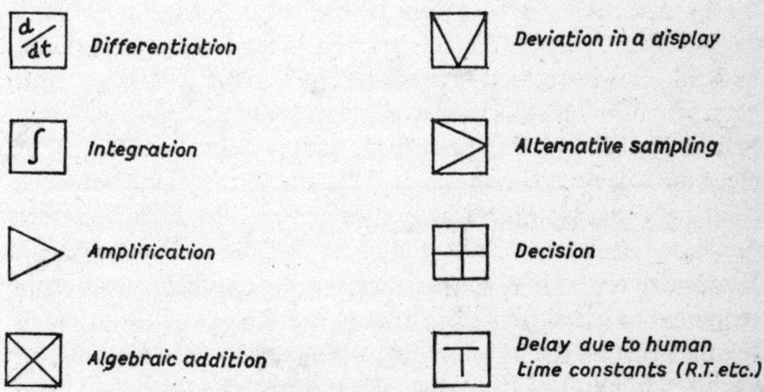

Fig. 2.2 *Symbols used in figures which follow (from Murrell, 1965a)*

series of delays between the making of the control movement and the ensuing display of the effect of the change. In systems with lags of this kind, there is a tendency to hunt (i.e. oscillate on either side of a desired value) which can only be prevented by the operator learning to anticipate the course of events and to make corrections before an indication appears that these are needed. This means that he may have to differentiate and/or integrate several times over and this we know he is very bad at doing. (Integration is the process of derivation of rate or change of rate of a function; differentiation is the opposite process and also means estimation of displacement from rate or acceleration.) The solution with this kind of situation is, as far as possible, to take the integration/differentiation functions away from the man and incorporate them in the machine and so reduce the task to a basic controlling function. But even a task as simple as this has become is not necessarily without its difficulties.

One simple task which has been extensively studied in the

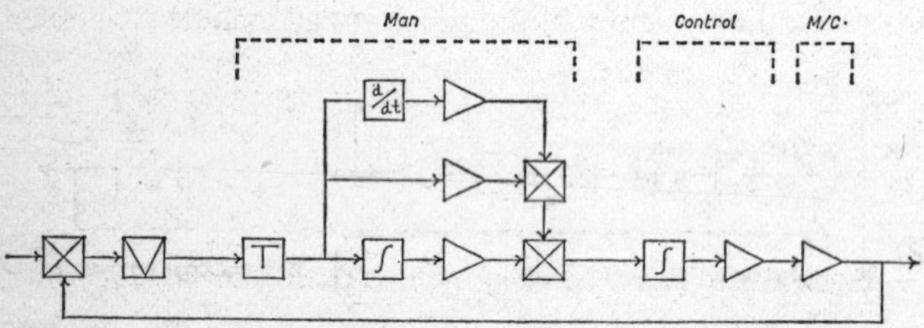

Fig. 2.3 *An 'unaided' tracking system (from Murrell, 1965a)*

laboratory is that of tracking where a subject gets an immediate response to his actions, but this is not always as simple as it seems because the target will be constantly accelerating and decelerating and so the operator's task may be described as in Fig. 2.3, which shows that he is expected to respond to a deviation by the operation of a damped joy-stick which will provide one integration with the machine acting as a simple amplifier. In this situation the man will have to act as a differentiator, an amplifier, an integrator and two adders at one and the same time, and this he will not do at all well. There is ample evidence to show that a man will do best if he functions as a simple amplifier so that if he tracks with perfect aiding the mechanism would be required to carry out integrations, differentiations and additions as is shown in Fig. 2.4. In other words the functions of the man and the machine have been reversed.

This assumes immediate feedback, but if a delay is introduced in this feedback, which is represented by the horizontal line along the bottom of the diagram, the situation becomes almost impossible. In order to overcome this difficulty, Birmingham and Taylor (1954) have proposed that the system should be 'quickened'. Quickening is designed to give the operator immediate knowledge of the result of any control movement which he may make; this is achieved by providing feed-forward

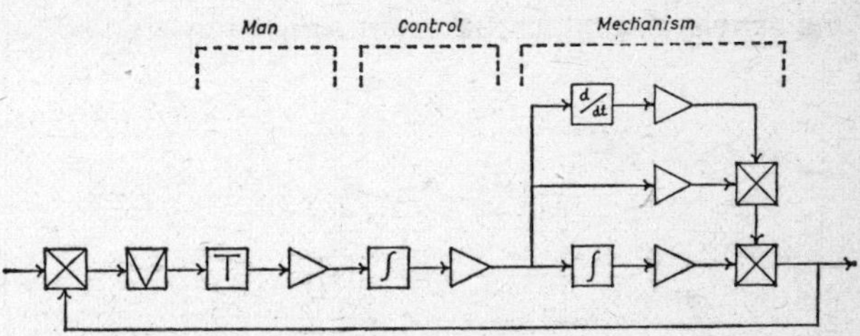

Fig. 2.4 *An 'aided' tracking system (from Murrell, 1965a)*

loops which may give information about position, rate, acceleration, or rate of change of acceleration to output integrators which give the operator immediate knowledge of the effect of his control action. The application of quickening to a simple tracking system is shown in Fig. 2.5. In an arrangement like this it is possible for the display to be a dial with two pointers, one of which gives the input while the other gives the output controlled by the quickened feedback. The task of the operator is simply to keep the two pointers in alignment at all times and Birmingham and Taylor have shown that hunting is virtually eliminated by such an arrangement.

Taylor (1957) reports an experiment in which subjects were required to track under four different conditions, three of which incorporated delays. In one of these, tracking was in one coordinate (the movement along a line formed by the intersection of two planes), and in two others in two coordinates, all without quickening; in the fourth condition subjects tracked in four coordinates simultaneously with quickening. Taylor's results showed that some subjects were able to cope with the first condition after eight days' practice, but that the second and third conditions were impossible; in the fourth, quickened condition, all subjects achieved perfect performance on the

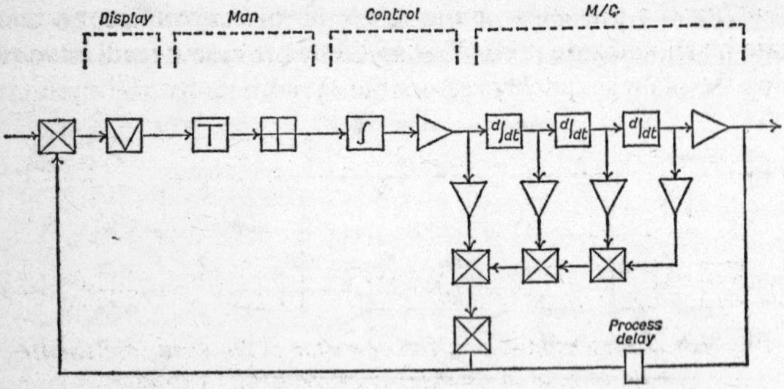

Fig. 2.5 *A 'quickened' system (from Murrell, 1965a)*

second day! From this Taylor concluded that quickening a display to give immediate knowledge of results not only increased the number of control loops which one operator could handle (in this experiment from 1 to 4) but also substantially reduced learning time. The kind of situations to which this technique is applied include operation of helicopters, particularly for blind night hovering, the steering of ships or the maintenance of depth in submarines, or the control of chemical plants where adjustments have to be made from time to time to the set points of the automatic controllers in order to compensate for changes in the raw material input. Without quickening there can be substantial and expensive hunting on either side of the ideal thru-put because of the length of time that must elapse between the operator making the adjustments and then learning what the effect has been. In some plants this delay can be as much as one hour and depends on a man on a bicycle pedalling like a bat out of hell between the plant and the laboratory.

Although in many of the situations quoted the operator may actually be taking action intermittently, the system is essentially a closed loop one. There are other situations, however, in which the task of the operator may be primarily to deal with deviations if and when they occur; in particular he may have to decide whether the deviation is likely to be temporary or whether the process is going to go out of control. In the first case he should ride it out, but in the latter case he will have to stop the plant as quickly as possible. Traditionally, the operator

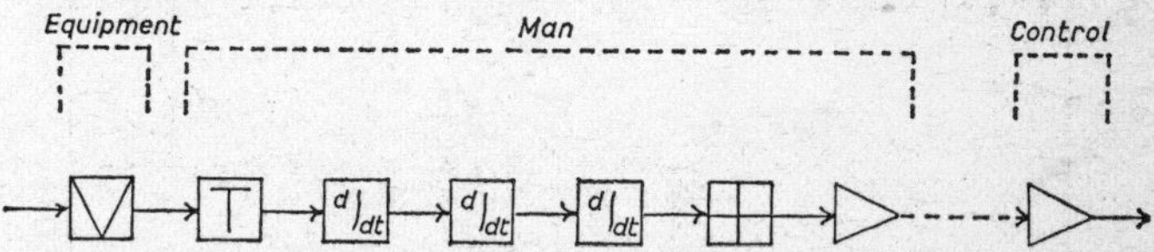

Fig. 2.6 *A system in which the operator is required to estimate rate of rate of change (from Murrell, 1965a)*

would be given an instrument such as a pressure gauge or a thermometer which would simply begin to show a deviation, and from the rate at which the needle might be moving it would be necessary for him to estimate the third time-differential, that is the rate at which the rate of change is accelerating. This situation is shown in Fig. 2.6 and is a task which an operator will find virtually impossible to do. In fact, he might just as well toss a coin and say 'heads I ride it out, tails I close down the plant'. In practice he would almost certainly close the plant down (with a consequent substantial loss of production), this being the safest decision; if he were to ride it out when he should have shut down the consequences might be obvious to everyone in the neighbourhood if a pressure plate blew off and the contents of a vessel were distributed over the surrounding countryside. As is shown in Fig. 2.7, the situation is easily resolved by transferring the triple differentiation function to the equipment and leaving the man simply to make decisions.

This example is taken from a real-life situation. The firm involved had built a large and expensive control room in the fond belief that their operator could make the required decision with the aid of a pressure gauge. The situation was made more difficult by the nature of the process: if the plant was shut down the product would solidify and it could take up to 36 hours to steam it out and get the process running again; if the operator decided to ride it out when he ought to have closed down, pressure plates would rupture and the product would be spilled

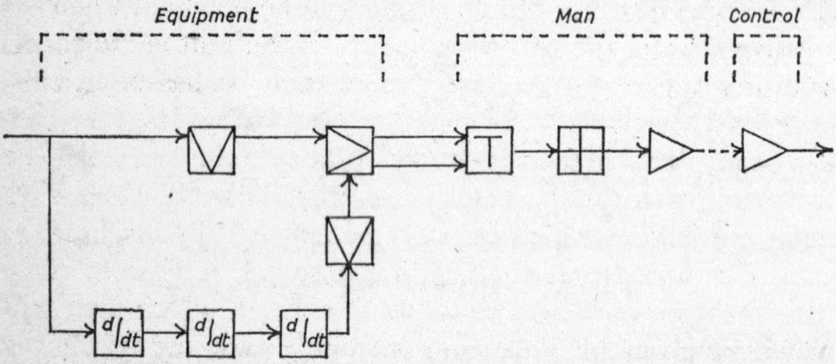

Fig. 2.7 *A system in which the functions carried out by the operator in* Fig. 2.6 *are transferred to the equipment (from Murrell, 1965a)*

all over the plant – another long cleaning-up job. So the operator invariably chose to shut down, thus rendering his job meaningless. If the system is redesigned there is no decision to be made, and a machine can take the necessary action. If the rate of rate of change is accelerating the plant must be shut down quickly, if it is decelerating it will be allowed to run on. Had the designers thought things out, a lot of very expensive, redundant control room equipment would not have been installed.

It is now becoming increasingly common for rate of change, rate of rate of change etc. of a displacement to be displayed to an operator whose task is thereby made simpler, since it is clear from a fair body of research that as a general principle the more closely the information given to the operator relates to that which the operator actually needs to know in order to take any action the quicker will be the learning process and the less will be the likelihood of mistakes being made.

Summary

In this chapter we have looked, albeit briefly, at the ways in which man and machine can complement each other and at some of the research which has led to the conclusions offered. We have seen that optimum results can be obtained when the relative roles of the two components of the unit are allocated with due regard to the capacities of each. It has often been suggested, particularly by computer manufacturers, that men can be displaced in most functions. While this of course may be theoretically true, it must always be remembered that men have many capabilities which can only be matched by machines at enormous expense, if at all. So for a very long time to come man–machine units will continue to exist in which man will play an important and usually controlling part, and it behoves us to ensure that the design of the hardware part of the unit is such that it can be operated with maximum efficiency. Details of some of the research findings which enable this to be done will be discussed in the chapters which immediately follow.

3
Machine to man

In order to carry out his controlling function man must know what his machine is doing – the machine must 'communicate' with him, if you like. The jargon word for this function is to 'display', so a display is the means by which a machine gives information to an operator, and in the strictest sense of the word this can be any device or event which gives information about a situation which is occurring or has occurred. In this sense the movement of the tool and the colour and nature of the swarf (a thin sliver of metal produced by a tool when cutting) on a centre lathe will tell the turner how his job is progressing. He can also get information from the sound which the tool is making while it is cutting. These are therefore visual and auditory displays. In this chapter, however, we shall be using the term in a rather stricter sense which Murrell (1965a) has defined 'as devices which give information about an event or situation where this information is not (or cannot be) obtained by seeing or hearing the event or situation directly'. Even this definition is a shade wide, so we will be confining ourselves to instrumental displays which give information about conditions in a piece of equipment or in a system; this would exclude information conveyed by the hand signals of a tick-tack man at the race course which might be included in the wider definition.

Instrumental displays are mainly of two types, visual and auditory. The former are most commonly used and are the most versatile. They include *inter alia* instruments in which the information is given by means of an index, by means of lights, by means of shutters or by means of letters or numerals. Instruments which give information by means of an index are commonly called dials and these even now are only slowly being replaced by alternatives.

Up to the Second World War, the design of dial faces and indices had been largely at the personal whim of the engineer involved. They were generally large, with a scale covering approximately 270 degrees and with extensive markings. The pointers were skeletal with flowery scrolls as counter-balances. Alternatively, scales covering only a segment were widely used; less common were moving scale instruments such as in the speedometer of some early Morris motorcars.

In all these designs it seems to have been assumed that the only information required by an observer was to be able to read an exact value. It is not clear who first pointed out that this was not necessarily the case; that dials were used for a variety of purposes for which the design requirements differed. Chapanis *et al.* (1949) made this point: 'there is a general principle we should keep in mind: a dial should not supply more information than the operator needs or can use'. They point out that in addition to their *quantitative* function, dials are very frequently used just to check, that is to say in a *qualitative* manner. Slightly later Murrell (1951) identified nine different uses, five without associated controls and four with controls. These have been set out more fully in Murrell (1965a). The most important uses are: for *quantitative* reading where the operator has to read the value shown by the display; for *check reading* where the operator has to notice deviations from a given normal value; for *qualitative* reading where the operator has to understand the meaning of a change of value; for *check controlling* where the operator has to use a control to restore a display to a given value whenever a deviation from this value is observed. Other uses are for *comparison*, for *warning*, for *setting*, for *tracking* and for *indicating*.

The results of a survey into the use made of dials was carried out by McCarthy in H.M. ships; no similar survey of uses in industry has been traced, but there is no reason to suppose that the results would be any different. The results of this survey are discussed in Murrell (1952) and are shown in Fig. 3.1. In

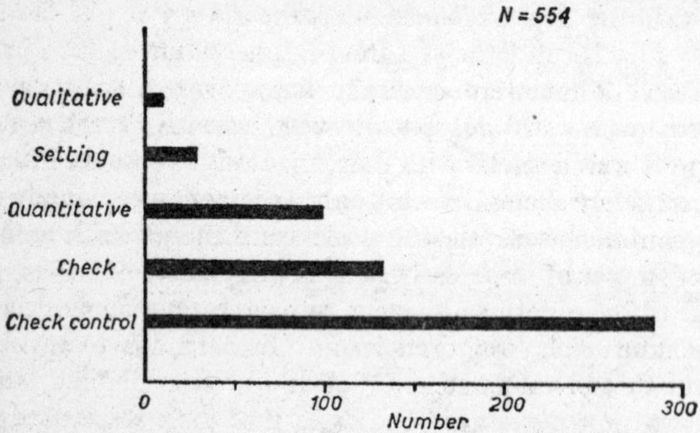

Fig. 3.1 *Results of a survey of the uses to which instruments are put (from Murrell, 1965a)*

spite of the fact that all the instruments were designed for quantitative reading, having normal numbered scales, some 75 per cent of them were used for some kind of checking and only 18 per cent for the purpose for which they were designed. It is true that some of the quantitative instruments had been in effect converted into check instruments, with the limits within which the pointer must be maintained being put on the cover glass in chinagraph pencil. It is quite clear from this survey that no real thought had been given either by the instrument designers or by the designers of the ship compartments to the use to which the instruments were to be put or to whether they were designed to give only the information which the operator needed to know.

We will now review some of the research which has been done into the design of the principal types of instrument. Though in actual fact the research was carried out piecemeal, followed no logical order and extended over a number of years, for reasons of clarity we will review these studies in some logical sequence.

Quantitative displays

Research on quantitative displays has covered, for dials, their best shape, the size and spacing of the scale marks, the design of the numerals and other legends, the design of the pointer and whether they should be white on black or black on white. The design of the bezels (the metal edge of the case which holds the glass in place) has also been investigated. Alpha-numeric (letters plus numbers) displays, which are rapidly superseding graduated dials for quantitative reading, have also been thoroughly investigated.

Dial shape

Almost the first, and certainly the most widely quoted, investigation into dial shape was carried out by Sleight (1948). He used a tachistoscopic display of five dial shapes (Fig. 3.2) for 0·12 sec. On the basis of the number of errors made, which ranged from 0·05 to 35·5 per cent, the shapes were ranked in the following order: open-window, semicircular, round, horizontal, vertical. The open-window instrument had a fixed index and moving scale; these results thus conflicted with other work (Christensen, 1952) which suggested that moving pointer instruments were superior to moving scale instruments. It has been suggested that Sleight's results could be explained by the amount of visual search required since the rank order is directly related to the physical size of the displayed scale. His technique has also been criticized on the grounds that the exposure time was about one-tenth of the time which would be taken if observers were allowed as much time as they required. In addition to comparing moving pointers with moving scales, in his experiment Christensen considered five other variables: the subject, the exposure time, practice, clockwise versus counter-clockwise and pointer fixation. All these variables were

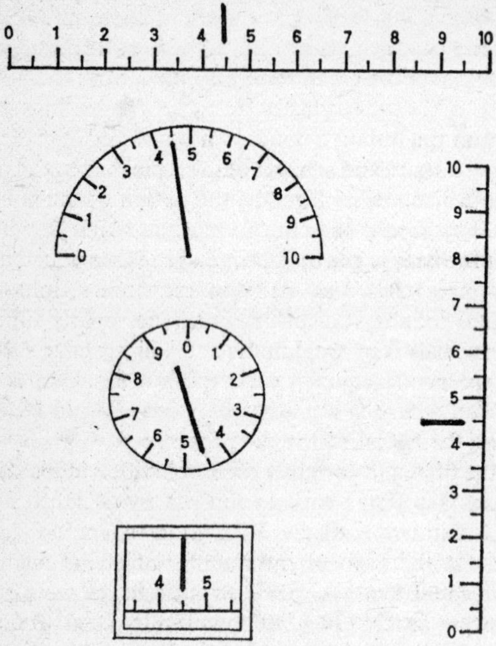

Fig. 3.2 *Five dial shapes tested by Sleight (1948)* *(from Murrell, 1965a)*

significant either singly and/or in combination with other variables but exposure time was found to be the most powerful factor. Some relationships significant at one exposure time became insignificant or even reversed at other exposure times, all of which suggests that the practice of inducing observers to read instruments faster than they would do naturally may well produce unrealistic results.

Kappanf *et al.* (1947) criticized Sleight's tachistoscopic method: 'the natural situation in dial reading is for most readings to be paced by the subject himself. Even when readings have to be hurried in accordance with the demands for the task, the dial reader is normally free to check himself by a second glance if he feels it is important to do so'. The alternative method which they evolved was to expose panels of twelve instruments at one time and to permit the subject to read these at his own speed and report the readings verbally. Each subject was timed from the end of reporting the reading on the first instrument to the end of reporting the reading on the eleventh instrument. Thus a time covering ten readings was obtained. This method has been criticized by Naylor (1954) on the grounds that it is possible or even probable that the subject moved his eyes to the next dial and may have started to locate the pointer while he was still speaking. He attempted to overcome this criticism by making his subjects switch on a light in order to observe the display and switch it off again as soon as they decided what the reading was. According to Naylor this would give the actual comprehension time.

Unfortunately, Naylor gives none of his data in his published paper, so it is impossible to see the extent to which his times differed from those of others. He gives results which he claims show that a straight instrument is more easily read than circular instruments and that horizontal straight scales are superior to vertical ones. In this he differs radically from Sleight who found that round scales were markedly superior to straight scales, but agrees with Graham (1956) and Thomas (1957), both of whom showed that horizontal scales were superior to round but that round were superior to vertical. Graham found that she had to use exposure time of not more than 0·5 sec in order to get any difference at all, while Thomas used the same exposure time as Graham and four shorter times down to 0·02 sec. The shape of the instruments he used were the same as that used by Sleight and the rank order of all were unchanged except for the open window, which at the slower speeds came second to horizontal whereas at the faster speeds it was the worst.

These confusing results have been made rather worse by Elkin (1959), who again used the exposure time of 0·12 sec on three types of dial – open-window, round and vertical; he also allowed subjects to make readings in their own time as proposed by Naylor. He obtained the same rank order as Sleight, with open-window the best and vertical the worst, but reports a much larger range of error than Sleight. For Sleight reported 0·05 per cent errors for the open-window whereas Elkin reported no less than 10 per cent at the same exposure time, and even when the subjects read the scale in their own time Elkin reported 0·85 per cent errors. The same discrepancies are to be found on the other two shapes.

And there the matter seems to have rested – inconclusively; in a sense this work has been overtaken by events since many quantitative instruments are being replaced by alpha-numeric displays. Furthermore, except in mechanically driven instruments such as speedometers, the torque, available from a bordon tube in the case of pressure gauges or a moving coil, say, in the case of electrical instruments, is insufficient to rotate the disc required in a moving scale instrument unless it is very light indeed and perfectly balanced and this causes substantial manufacturing difficulties. Nevertheless, the research is interesting in that it shows the different methods which can be adopted by psychologists working on a comparatively simple problem and the diversity of results which can be obtained.

Scale design and dial size

This work on dial shape is instructive since the psychologists involved could not agree either on the method to be used nor on the outcome. This is equally true of the research into the simple problem of scale design – how many marks should there be and how far apart? Research workers in several countries, using various methods, have come up with a variety of differing answers, one of which has been accepted as the basis of a national standard. This research will be given in some detail as an illustration of how experimental psychologists are unable to agree on either the method of research or on its outcome.

Until comparatively recently it was always assumed that the larger the dial face and the more marks which were packed into the scale, the more accurately would the instrument be read. And this often led to the absurdity that a scale could be read accurately to $\pm 0{\cdot}1$ lb whereas the intrinsic accuracy of the mechanism was probably no greater than $\pm 1{\cdot}0$ lb. In practice instruments were made to a dial diameter which was convenient for the manufacturer and the actual length of the scale which could be contained within the specified diameter was not considered to be a matter of very great importance. In aircraft, however, dials were required to be small and when research into instrument design was started for the US Air Force, the researchers concentrated more on the scale and let the dial size be consequential. Apparently, some of the earliest work was undertaken by Loucks (1944) who investigated three different designs of tachometer scales. The scales went from 0–35 and major scale marks were numbered every five units. One scale contained thirty six marks (the extra one being for zero). The second scale had 71 marks while the third scale had 176 marks; there were, therefore, 175 scale divisions, each worth 0·2 unit which was the accuracy with which the scales were required to be read. The method was tachistoscopic (i.e. exposure for a brief interval of time) at 0·75 sec. Loucks (1944) found that it was the most open scale which was read most accurately – 34 per cent errors – compared with 41 and 44 per cent errors for the other two. This means that a scale division representing $5 \times 0{\cdot}2$ units but which did not have any scale marks in it was read more accurately than one that did have a scale mark for every unit. In other words, the observers were successfully interpolating a scale division into five parts.

This idea was followed up by Kappauf *et al.* (1947) using the method which has already been described. They used a 360 degree scale covering 100 units with three different scale designs, one with a mark every unit (100 × 1), a second with a scale mark every five units (100 × 5), and a third with a scale mark every ten units (100 × 10) (Fig. 3.4). They also worked

Fig. 3.3 *These are three tachometer dials studied by Loucks. Taking short glances at each dial, observers made fewest errors with the one on the left (after Loucks, 1944) (from Chapanis, Garner and Morgan, 1947)*

with three dial sizes of 0·7, 1·4, and 2·8 in diameter at a reading distance of 28 in. (Imperial units were used throughout the research reviewed. Metric equivalents will not be included.) They worked under both 'speed and accuracy' instructions and 'accuracy' instructions and found that, in general, subjects

Fig. 3.4 *Three scales compared for legibility by Kappauf, Smith and Bray (1947) (from Murrell, 1965a)*

working under the latter instructions produced fewest errors. For all three scale designs the largest dial size was read the most accurately; under accuracy instructions there was very little difference between the 100 × 1 and the 100 × 5 at all dial sizes but 100 × 10 was much less accurate. The error rate when a tachistoscopic method was not used was very much smaller, varying between about 2 and 8 per cent for the three largest dial sizes. The reading time was between 2·0 and 3·5 sec under the accuracy instructions and 1·4 and 2·0 sec under speed instructions.

A similar experiment was carried out by Grether and Williams (1947), who used four dial sizes between 1 and 4 in. read at a distance of 30 in. Each scale ran from 0–50 units with six graduation marks so that interpolation was in tens. They also used four different distances of separation between the scale marks, making sixteen different designs in all. Their results are presented in terms of errors related to the physical distance between two scale marks and show that errors become minimal when this distance is 0.5 in. Similar results were obtained by Kappauf and Smith (1948), who came up with a figure of 0·44 ins for scale marks 10 units apart. They also showed that there were fewer errors when interpolation was into fifths, and this result has been confirmed by Leyzorek (1949).

It should be remembered that all this work was done by people who were interested in aviation instruments at distances between 28 and 30 in. from the eye of the observer. It seemed quite proper to them simply to specify the physical amount of separation between two scale marks for reading at this distance, and it did not seem to have occurred to them to extend the research any further. However, as Murrell (1965a) has pointed out in describing the logic behind work in the UK, it is not unreasonable to assume that if the distance between two marks is 0·5 in. at 30 in. then it should be 1 in. at 5 ft and 2 in. at 10 ft and that this would produce dials which were far larger than those currently in use, large though these might be. What is of course common to all these sizes is that the visual angle subtended at the eye by two scale marks is the same for each combination of size and distance. For example, at 20 ft reading distance the corresponding dial diameter would be 20 in.

Because of the absurdly large dial sizes obtained by extrapolation and because no one had investigated dial reading at any other than close distances, the staff of the Naval Motion Study Unit (NMSU) in the Admiralty undertook extensive experiments in 1951. Using six naval chief petty officers as subjects they tested five dial sizes and six reading distances. The dials ranged from 1·6–6·6 in. (41–168 mm), and the reading distances from 2·5–24 ft (760 mm–7·3 m) so that different combinations of dial size and reading distance gave similar visual angles. The same scale was used throughout; it covered 270 degrees and had a range of 0–200 units. This maximum scale value (msv) of 200 was chosen because the experimenters were advised by pressure gauge manufacturers that the intrinsic accuracy of the mechanism of 'commercial' pressure gauges was 1 per cent. This was interpreted as meaning that a reading which was plus or minus half a per cent about the 'true' value would be considered to be correct and this could be achieved on a 0–200 scale without the observer having to depart from whole numbers. For instance, if the 'true' value was 98, then a report of 97 or 99 would be considered to be correct.

There are a number of different ways in which a scale from 0–200 can be graduated; and so a preliminary experiment was carried out in order to determine which layout and numbering system would be used. The scale chosen had major numbered scale marks every 20 units with an unnumbered minor mark every ten units, so that by interpolation into fifths a scale should be read to the required accuracy. Subjects were told the accuracy with which they were required to read the instruments and they were instructed to read them as accurately and as quickly as possible. They were therefore performing under 'speed and accuracy conditions'. The experimental method was that of Kappauf *et al.* (1947) with boards of twelve dials for each size being exposed to each subject at each distance during each session. As each subject did fourteen sessions, a total of 5,040 readings were made by each subject. It became clear immediately that there was a very marked learning effect on this task so the actual scoring was confined to the last half of the experiment, that is the final 2,500 trials.

The outcome of the experiment was first published in 1951 in the Naval Motion Study Unit Report No 47, and later appeared as Murrell *et al.* (1958). Now when the data were considered, it rapidly became clear that the fundamental unit in dial reading is the angle subtended at the eye by that part into which a scale division has to be interpolated if the scale is to be read to the required accuracy. This is now known as an interpolated division and the terms used are those defined in BS 3693 of the British Standards Institution. If a scale is being interpolated into fifths, then the angle just referred to is that subtended by one-fifth the space between two scale marks. As will be seen in Fig. 3.5, accuracy is very close to 100 per cent down to a visual angle of 2 mins of arc; below this level accuracy falls off rapidly and linearly with the log of the visual angle. The results from Kappauf and Smith (1948) and Grether and Williams (1947) can be plotted in much the same way, but the point at which a rapid decline in accuracy commences lies somewhere between 5 and 9 mins of arc. This is what is to be expected, since instruments designed according to their data are going to be very large, as has already been mentioned.

The reason for this major discrepancy between American and British work is not difficult to find. The Americans appear to have used as subjects students who were unfamiliar with the task of dial reading, and the number of trials with each design of scale was quite small. For instance, Kappauf and Smith (1948) gave their subjects only sixty trials on each of eight designs, and Grether and Williams (1947) gave their eighty subjects only about eighty trials on each of four dial designs. By contrast, the British work used naval chief petty officers whose regular job involved dial reading and more than 2,500 trials were used for practice before recording started. There was another difference: in the American work different dial designs involving different accuracy of interpolation were used in the same experiment, whereas in the British work the same design and interpolation requirement was used throughout. It can be argued that in practice observers will move from one dial to another and that adjacent scales will not be the same. This may well be true, but if the instruments have been properly designed they should differ only in the maximum scale value

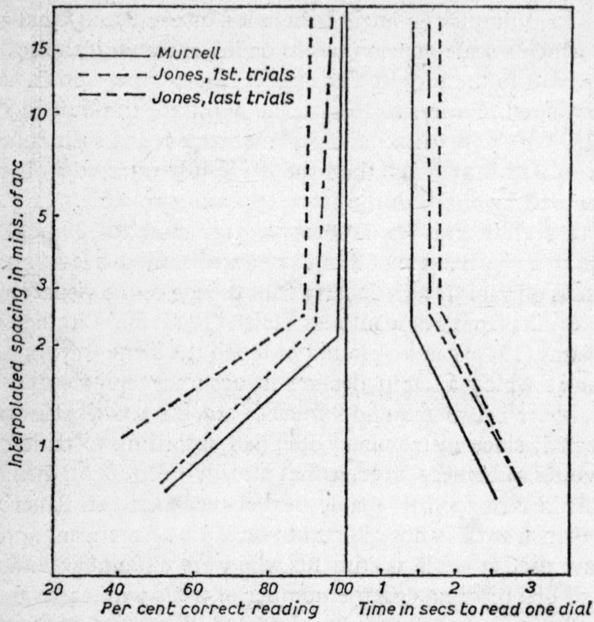

Fig. 3.5 *The relationship between reading accuracy, reading time and angle subtended at the eye by an interpolated spacing* (*after Murrell* et al., *1958*)

and the requirements of interpolation should be the same throughout. Moreover, in all the work which has been undertaken, it seems to be an implicit assumption that the people using the resulting dials will use them regularly; they will be pilots in the case of the American work and naval personnel in the case of the British work. It thus seems more reasonable to base the design on results from practised subjects than on data obtained from subjects who were very naïve. In spite of this, the recommendation that for viewing distance of 28–30 in. the interpolated spacing should be 0·07 in still persists (Morgan *et al.*, 1963). Applying this recommendation to a 0–100, 270 degree scale the diameter of the dial blank would be 3·7 ins, which is substantially larger than a recommendation based on the British work.

Two further points should be made. First, Churchill (1959), on the basis of some tachistoscopic research, questioned the so-called 'law of constant visual angle'. However, his technique and scales differ so much from work which has previously been reported that it is difficult to accept this suggestion.

Secondly, the experiment of Murrell *et al.* (1958) was later repeated by Jones *et al.* (1965), using smaller dials and even closer reading distances than did Murrell or the American workers. Their results confirmed exactly those of Murrell *et al.* that the optimal interpolated division was 2 mins of arc ($\frac{1}{30}$ of a degree). It is to be noted that the results of Jones *et al.* not only cover the range of sizes and distances used by the American workers but also refute Churchill's suggestion that 'the law of constant visual angle' does not hold.

The production of a British Standard

All this was to be in the future when the British Standards Institution (BSI) towards the end of 1951 decided to set up a committee to produce a standard on the graduation of instruments. One of the first things that this committee had to decide was which research it had to accept. So far as the relationship between dial size and reading distance was concerned, the committee decided to accept British research results on the grounds that they were more realistic and accorded more closely with practice. Not unreasonably they wished to be sure that what they were doing was correct, so they commissioned several additional studies as the work proceeded. One of the first of these studies was to determine the optimum method of graduating a 0–200 scale. There are five different ways in which this can be done (Fig. 3.6). This had already been investigated by the NMSU researchers as a preliminary to their main experiment, but it was felt that a fuller experiment would give guidance on the optimum number of numbered scale marks and the numbering system which could be extrapolated to scales with other msv. The work was undertaken by John Spencer, a psychologist then employed by the National Coal

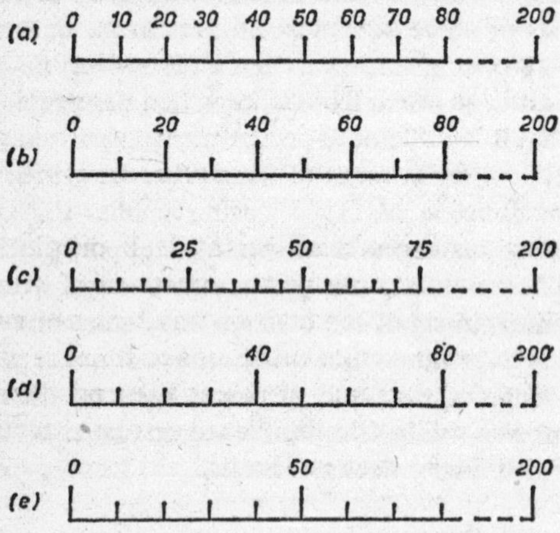

Fig. 3.6 *The various ways in which a 200-unit scale could be numbered (after Murrell, 1952)*

Board, and Mrs A. Dobbie, who reported in May 1954 to the BSI that a scale with the major marks numbered in 20s (that is, the scale used by Murrell *et al.*, 1958 in their experiment) had proved to be superior to other designs tested.

Before the proposed standard could be finalized, it was felt necessary to ensure that the recommendations would produce a scale which could be read by typical users. So Spencer took what were perhaps the most awkward scales – 0–6 and 0–600 – and used, as subjects, twenty-four process operators ranging in age from 21–50 years employed at the British Petroleum Refinery at Llandarcy. On average they did not achieve the accuracy of 95 per cent of readings within plus or minus 1 per cent of msv which was envisaged by the BSI, but they did only 100 readings on each dial design. Even so, there was a very marked improvement over this short learning period, with performances of the order of 90 per cent being achieved (Spencer, 1961).

Eventually, in 1964, BS 3693, part 1 appeared dealing with 'commercial' instruments read 'at a glance', that is instruments with an intrinsic accuracy no better than 1 per cent resolution. Work immediately commenced on part 2, which brought the intrinsic accuracy down to 0·33 per cent resolution. Part 2 was published in 1969 and although work had proceeded quite far on the preparation of part 3, which dealt with really accurate instruments, this section has never appeared and the committee seems to have died.

During the time this committee was deliberating, there were never less than two psychologists participating, with Spencer in particular producing a number of design memoranda on such matters as scale reading accuracy and numeral design and the meaning of such terms as 'observation error'. Thus it could properly be claimed that BS 3693 is the first British Standard to have been produced on the basis of evidence produced by experimental psychologists (Murrell, 1973).

One thing that became apparent throughout the work of the BSI Committee was the number of gaps in the detailed knowledge required to produce a series of firm recommendations. For instance, when the International Standards Organization (ISO) became interested in this area and organized a conference in 1969, the French claimed that in certain circumstances it was better to interpolate into fourths rather than fifths. They did not produce any experimental evidence for this but claimed that it was a result of 100 years of experience! Thus if a 0–10 scale is to be read to the nearest half unit, which requires that there should be twenty interpolation intervals in the scale, there can either be three numbered minor marks with interpolation into fifths or four numbered minor marks with interpolation into fourths (Fig. 3.7). Now the numbering system for interpolation into fifths is in units of 2·5, which is considered to be unsatisfactory and certainly less satisfactory than numbering in twos, which is required for interpolation into fourths. Since there was no evidence to suggest that interpolation into fourths under these particular circumstances would be less satisfactory than interpolation into fifths and because the number system was superior, the French proposal was accepted by a British delegation. But it seemed desirable that interpolation into fourths should be investigated, so a study was carried out in the

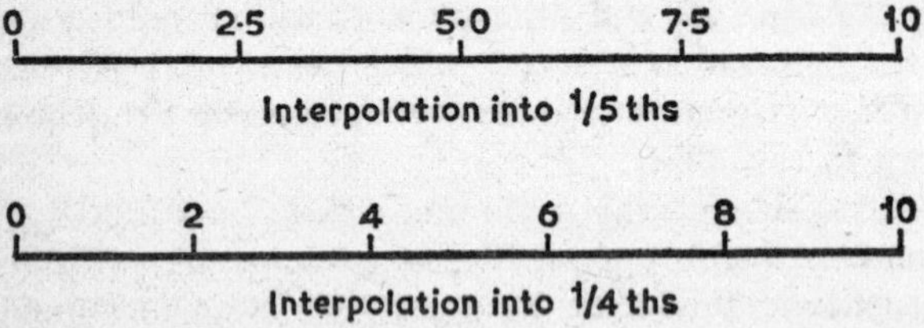

Fig. 3.7 *The effect of scale numbering on interpolation*

Department of Applied Psychology, UWIST, in the summer of 1970, which showed that interpolation into fourths was actually superior to interpolation into fifths for the particular circumstances which have been described (Murrell, 1973).

This work has been quoted at some length because it shows how psychological research can be brought to fruition and have a marked influence on the design of one part of the man-machine interface. It also shows that, if one is to judge by the writings of psychologists, what are believed to be adequate design data prove to be full of gaps when practical people get down with psychologists to solving real problems. Finally, this work illustrates the big differences which may be found between the results of different groups of research workers. Differences which are totally incompatible and which certainly require explanation have been explained (see p. 47). The dimension for the interpolation interval produced by the Americans is very much larger than that produced by the British. The former work was done more than twenty years ago and no attempt seems to have been made to verify it since, yet in spite of the success of instruments design to British Standard, not only in Europe but also in America, the findings of Kappauf, Grether and Williams in the late fifties are still uncritically quoted in American Human Factors Engineering textbooks. In this instance, the differences can be explained on the basis of the amount of practice given to subjects before recording started. But there is no similar explanation for the differences between the views of the American and British researchers and that of the Canadian, Churchill, on the so-called 'law of constant visual angle'; all the evidence seems to suggest that it does hold good, as also does practical experience of the working of BS 3693.

Scale base length

Accepting, then, that the optimum interpolated division should be 2 mins of arc, then the distance along the scale base corresponding to an interpolated division will be 0·08 in. at a reading distance of 12 ft. Now this is for observers with 6/6 Snellen eyesight; to allow for people with less good eyesight, say down to 6/12, this value could be increased by about one-quarter to 0·10 in. Commercial instruments with an intrinsic accuracy of one per cent would therefore be required to be read to one-hundredth part of the msv, so there will be 100 interpolated divisions in the optimally designed scale, that is the scale-based length should be 10 ins for a reading distance of 12 ft. If the same units are used for both dimensions, then

$$D = 14{\cdot}4\ L$$

where D = the reading distance and L is the scale base length,

or

$$L = 0{\cdot}0694\ D$$

This is not the end of the story because the scale must obviously be inscribed on to something, and if the instrument is circular it is likely that the scale will be wrapped round 270° to give a scale base diameter of 0·0295 D. This in turn will, as a matter of common practice, be $\frac{5}{6}$ of the visible area of the dial blank which will therefore be 0·0354 D. Thus when the reading distance is 144 in. (12 ft) the diameter of the visible area of the dial blank will be 5·10 ins, which would certainly be accommodated in a case, with appropriate bezels, of 5·5 in. diameter. It is unlikely that most manufacturers make instruments this size; the standard is much more likely to be 6 in. and so it is this that will be chosen should it be required to be read from a reading distance of 12 ft. You can of course work backwards from a standard case size and specify the maximum reading distance for which the instrument is suitable; this practice has for some years been adopted by some manufacturers.

Other design considerations

So far all we know about a scale is that optimally it should contain 100 interpolated divisions and that there is an optimum reading distance for any particular instrument size. But before a scale design is completed, it would be necessary to know how many scale marks there should be, what system of numbering will be adopted and how many numbered major marks there will be; what will be the dimensions of these marks and will there be 'tramlines'; what should be the shape and size of the numerals and should they be inside or outside the scale; what should be the shape of the index (pointer); in which direction should the indicated value increase; what should be the colour of the scale marks, the dial face and the bezel?

Although psychologists have been involved in research into all these aspects of design, space will allow mention of only the barest outline of some of the most relevant material.

The number of scale marks and how they are numbered will depend upon the msv required. Over the years a number of msvs have been established by convention and for some of them it is virtually impossible to design a scale which conforms to the recommended criteria. If the intrinsic accuracy of the mechanism of a commercial instrument is one per cent it is generally accepted that you should be able to read the instrument to one per cent. So if interpolation is into fifths, the ideal scale should have twenty scale divisions. This can easily be achieved in a scale such as 0–100 or 0–200, but the scales 0–150 or 0–300 do not meet the requirement.

Digressing slightly for a moment, we find that the relative merits of the different ways of numbering the major marks have been investigated by a number of researchers (e.g. Vernon, 1946; Barbour and Garner, 1951), all of whom agree that numbering in ones (1,2,3 . . .) or twos (2,4,6 . . .) are the preferred systems, with fives (5,10,15 . . .) lying a very close second. All other numbering systems are found to be poor with 2·5s very bad indeed. Although this has not been shown experimentally, it seems reasonable to assume that people would tend to think in ones or twos or fives and that they would tend to read scales to one or other of these units, or to decimal multiples or sub-multiples thereof, even though a scale in theory can be read with greater accuracy. For instance, consider a 0–40 scale with twenty scale divisions. It is clearly possible to interpolate a division into fifths and read to the nearest ·4 of a unit, but it seems far more likely that each division would be interpolated into quarters so as to read to the nearest ·5 of a unit. This means that reading will be to 1·25 per cent of the maximum scale reading, a lesser accuracy than the instrument is designed to give. The converse can be illustrated by the 0–300 scale with thirty scale divisions, which is likely to be read to the nearest 2 units which is 0·7 per cent of the maximum scale value, a greater accuracy than the scale is designed to give.

From what has been said, it will be seen that the choice of the msv, the number of scale divisions and the way in which the major scale marks are to be numbered is largely a compromise and that in practice the scales are likely to be read with an accuracy less than the design accuracy. A full discussion of the practice which leads to the choice of the msv and the general layout of the scale marks will be found in Murrell (1965a).

The desirable length and thickness of scale marks has been the subject of a number of investigations, the results of which suggest that, within limits, some variation can be accepted without influencing the reading accuracy to any extent. It is clear that long marks are undesirable since they produce a palisade effect. It seems generally to be accepted that the minor marks should not be longer than the space between them and that the major marks should be twice the length of the minor marks. The thickness of the marks is a compromise between what can clearly be seen and too much encroachment into the space between the marks which is going to be interpolated. There is some evidence that when interpolating the observer does not work mentally from the centre line of the marks but from their inner edges.

In earlier days, when manufacturing methods were rather more crude than they are now, it was the custom to join the ends of the minor marks with continuous lines, which were commonly spoken of as 'tram lines', in order to cover up irregularities. The effect is to turn the scale divisions into boxes. No evidence has been published on whether these boundary lines cause difficulties or not, but it seems generally to have been assumed that it is easier to interpolate an open-ended space than a closed space. Opinions differ on whether the major marks should project inward and be numbered on the inside of the scale or outward and be numbered on the periphery; both factions have produced evidence to support their views. From the practical point of view, numbering on the outside of a circular scale means that you will have to have a larger case than would otherwise be necessary; on a segmental scale it has no effect and scales numbered on the outside are not infrequently to be seen.

For the numerals, there are three parameters all of which have been investigated by a substantial number of research workers: the height–stroke width ratio; the ratio of the width of the numeral to its height; the actual shape of the numeral. The researchers agree reasonably well on the first two but have come up with some quite odd shapes for the third. Not all the designs were specifically for dials but the general implication is that if the numerals are easy to read they can be used in most contexts. The British Standards Institution commissioned an alphabet for BS 3693, paying particular regard to the work of Soar (1955) into confusion between numerals (Fig. 3.8). The digit 5 was the most frequently confused – with 6 with 8 and with 9. Based on his findings Soar himself (1958) designed three sets of alphabets in which common elements were minimized and unique elements intensified, but they were too

Fig. 3.8 *Instrument digits recommended in BS 3693A: 1964 (by courtesy of the British Standards Institution)*

unusual to be acceptable. One advantage of the BSI design is that the type face Granby Bold can be used as a substitute.

Most of the research on numeral visibility can be criticized on the grounds that requiring subjects to read numerals presented randomly in a variety of ways differs from the real situation where numerals would always go in sequence. Thus a numeral laying between 4 and 6 is hardly likely to be called 8 or 9. It is clear that observers would become familiar with the numbering of a dial and Murrell *et al.* (1958) demonstrated that scales can be read with acceptable accuracy at a distance at which it is impossible to read the numerals when they are presented singly in random order. It is clear, therefore, that for practised observers the numerals cease to have any real importance nor, by implication, does the research on numeral design. On the other hand, in fairness to the research workers it must be pointed out that their objective was not by any means always the design of numerals for dials. It is the compilers of 'cook books' who must be blamed for quoting the results as if they had been intended for use on dial design. But even so, the research workers who set out to design numerals for dials appear to have overlooked in their laboratory technique the actual way in which the dials are read, and this illustrates once again the gap that can develop between psychologists in the laboratory and practical engineers in real life.

Very little research on pointer shape appears to have been published, and this has been summarized by Spencer (1963). Some work was underway in America in about 1948 and a progress report suggested that it was advisable to use a thin pointer with a fine taper terminating in a relatively blunt tip whose sides made an angle of approximately 90°. The tip should be approximately the same width as the thickness of the minor scale marks. White *et al.* (1953) tried the effect of enlarging the tails of the pointers when used for check reading. Their results were inconclusive. There was a suggestion, however, that fewer 180° reversal errors were made with a pointer without a tail. On the other hand the tail acts as a counter-balance to reduce the torque required of the mechanism, so where a tail is required, as it usually is, it should be the same colour as the dial face, while the remainder of the pointer from the collet to the tip should be in maximum contrast.

The clearance between the scale base and the tip of the pointer has been investigated by several research workers. Vernon (1946) concluded the tip should be as close as possible to the scale and Churchill (1956) suggested that clearance should be between 0 and 15 mins of arc visual angle. This research has assumed that the dials are circular and that the pointer is in the centre. Naylor (1954) showed that with segmental curved scales this arrangement was far superior to the arrangement in which a pointer is placed outside the scale.

This seems reasonable since there is more of the pointer to catch the eye, but Spencer (1963) found no difference on two 70° scales. Naylor also concluded that with vertical straight scales the pointer should be on the right-hand side of the scale with the numerals on the left and that with a horizontal scale, the pointer should be above the scale with the numerals below.

Maximal visual efficiency will be obtained when there is maximal contrast between the dial markings and the pointer and the dial face. This usually implies that one should be black and the other white. A number of experiments have been carried out in order to determine whether black marks on a white background are superior to white marks on a black background. Most experimenters have been interested primarily in aircraft instruments which are used for night flying, so recommendations that dial faces should be black and the numerals or markings should be white are common. Kuntz and Sleight (1950) and Naylor (1954) both report that there is no difference in visibility with a well-designed scale whether it is black on white or white on black. On the other hand, Papaloïzos (1961) reported that dials with light faces were read more accurately than were those with dark faces. There is reason to believe that visual acuity is impaired when the surroundings of task are lighter than the task itself (Lythgoe, 1932), from which it follows that for normal usage the face of a dial should be white.

The most important feature of a dial is its pointer. So it is important that the bezel or case should not conflict with the pointer by being black. It should be coloured just sufficiently to separate the dial from the panel and should never be highly polished. In certain circumstances it is useful to use the bezel for colour coding.

Features in the design of quantitative instruments have been dealt with at some length because it is an area which has engaged the interest of psychologists over a number of years and because in the United Kingdom the research has resulted in the production of a British Standard. It is indeed ironical that so much effort should have been put into a type of instrument which is relatively less used and which for many purposes is being replaced by a form of digital presentation. There has clearly been a misunderstanding in many people's minds, both users and some researchers, as to exactly what information an observer requires from an instrument, it generally being assumed that an instrument designed for quantitative reading is also suitable for check reading, usually with some sort of a limit marked on the dial face or the cover glass. As a result, instruments designed specifically for checking have received relatively little attention.

Test instruments

A type of quantitative instrument which will undoubtedly hold its own for a long time to come is the test instrument, which has an intrinsic accuracy greater than one per cent and which would normally be read by an observer who has plenty of time to be sure that his reading is correct. Most of the research outlined above was undertaken under speed and accuracy conditions, which are not the conditions under which a test instrument will be read. The applicability of the results to test instruments is therefore problematical. Presumably because they are not used in aircraft, test instruments apparently did not receive any attention from the American research workers and it wasn't until BSI started work on Part 2 of the standard that any research seems to have been undertaken.

James and Murrell (1967) confirmed that there were fewer errors and quicker learning under accuracy instructions when interpolation was into halves compared with interpolation into fifths, into tenths and no interpolation at all. They also confirmed that interpolation into fifths was superior to no interpolation and these were both in turn superior to interpolation in tenths. The criterion of accuracy of 95 per cent established for Part 1 of BS 3693 was shown to be conservative for interpolation into halves. Nevertheless, without any real supporting evidence BSI decided that down to an intrinsic accuracy of

0·33 per cent interpolation should continue to be into fifths, largely on the grounds of expediency and to avoid greatly overcrowded scales.

Instruments for check reading

Although check reading – checking whether the indicated value is within prescribed limits – is by far the commonest use of dials, the design for this purpose has been relatively little investigated. And that which has been done has been by way of converting quantitative instruments, usually circular instruments, for the purpose. The outcome of experiments on check reading appear to depend on whether a consequential control movement is required or not. Grether and Connell (1948) showed that when an up-and-down response is required the most satisfactory results are obtained when the 'safe area' is in the 9 o'clock position on the left-hand side of the dial face so that an upward movement of the pointer corresponds to an increase in value and a downward movement of the associated control would produce a downward movement of the pointer. However, Morley and Suffield (1951) have shown that where a verbal response is required there is relatively little difference between the different sectors of the circular dial. Quite early on it was argued, based on work on the speed of reading quantitative instruments already quoted, that the longer pointer on a circular dial will be seen more easily than the shorter pointer on a straight dial. However, this has not been borne out by experiment, since Murrell (1960b) has shown that deviations on vertical straight dials were noticed rather more quickly than those on circular dials. In his experiment the centres of all instruments were the same distance apart, but clearly vertical straight scales have an additional advantage in that they can be put closer together than can circular scales and therefore the distance to be scanned will be reduced and this may improve detection.

Only one design of instrument specifically for check reading has been traced. Kurke (1956) proposed that a pointer should be mounted on the centre line of a segment cut from a disc, the whole assembly rotating on the spindle of the instrument (Fig. 3.9). The face behind the disc would be covered in such a way that it would be exposed to give a high contrast wedge when a deviation had occurred, which by its position would

Fig. 3.9 *Design for a special check instrument proposed by Kurke (1956), showing* (right) *the 'pointer' and dial face and* (left) *safe and danger conditions. This instrument can also be used for quantitative reading (from Murrell, 1965a)*

clearly show whether the reading was too high or too low. He tested his design against two standard instruments, one of which had a safe area marked upon it; deviations were noticed in 4·3 sec as compared with 20·5 sec for the standard instrument.

It may be that the importance of a proper design of a check instrument has been somewhat over-emphasized. In practice in large control rooms all important systems are fitted with alarms which will go off when a function goes out of limits and this in turn can produce certain design problems to ensure that, where speed is of the essence, the operator understands the meaning of the flashing lights which are characteristic of this particular system. No psychological research on this aspect of machine–man communication has been traced. Where process operators are expected to monitor a function, it is common practice to use chart recorders which show very clearly whether an indicated function is approaching high or low limits; this is probably the best form of check instrument since it shows trends and the rate at which a function may be altering. However, the majority of check instruments are probably still used outside control rooms (the car speedometer is essentially a check instrument) and should thus receive more attention than they have.

Indicators and warning devices

An indicator gives non-quantitative information about the state of a system when it is under human control. Two main types can be distinguished – shutters and pointer type instruments. There appears to be no research on this type of visual display since the kind of information given is simple, such as on/off, open/shut and so on. Devices of this kind which are in use seem to have been evolved entirely by experience.

Warning devices are required to call the attention of the operator to some action which he has to take in relation to the equipment: in this sense the alarm systems described in the previous section are warning devices. In general the old type of shutter or 'bull's eye' such as were found on manual telephone exchanges have largely been superseded by warning lights, the main disadvantage of which is that the absence of a light may mean either that all is well or that electrical circuits have broken down! A system was evolved by NMSU in which a signal light burnt dimly all the time and changed to bright flashing when a warning had to be given. The alternative is to use two lights, one green or amber and one red, the former showing that all is well and the latter flashing to give a warning. Lights which go out to convey a warning have been shown to be quite useless; a light must flash.

The meaning of a warning can be conveyed either by the shape of a light or by its position, by its colour or by trans-illuminated labels. When seen singly it appears that not more than five coloured lights can be identified with confidence and then only if the colours are very carefully selected. Three- or four-colour systems can be used with complete confidence. Holmes (1941) has studied the chromaticity tolerances of warning lights and concludes that blue and purple are not satisfactory since, unless the blue is very saturated, it can become confused with green and white. Red is a good colour. In 1959 the International Commission on Illumination (CIE) published a report on the colours of light signals which give recommendations for three-, four- and five-colour signal systems. These are in the form of equations of the boundaries of the recommended colours. Due to the difference in the sensitivity of the eye to different colours, blue and purple in particular will have to be from fifteen to twenty times stronger than a white light in order to appear equally bright.

If the information is to be given by means of coloured lights, it is to be remembered that a proportion of the population is colour blind. This has caused a certain amount of difficulty in America where traffic lights are not always one above the other as they are in the United Kingdom. Experiments have been carried out with different shaped ribbing on the 'stop' and 'go' lights so that colour blind people can tell which light is on. This kind of solution would not be possible in industry so that where efficient operation of equipment depends primarily on distinguishing colours, a prospective operator must be tested for colour blindness.

Auditory displays

Simple auditory displays have been with us for a very long time. In old mansions a row of bells in the servants' quarters would inform which room required service. Slightly more sophisticated are the codes used in the telephone system which involve both pitch and temporal distribution of the sound to inform the caller when to dial, whether the number is engaged, when to pay or whether the number is unobtainable. It is not known, since research if any has not been published, whether these particular displays were chosen arbitrarily or as a result of some investigation; one suspects the former. More recently, sophisticated auditory displays have been developed for air traffic control.

Special displays

Up until now we have been looking at psychological research on the design of the most ubiquitous forms of display. But there are other, special, displays which have received attention. One of the most important is undoubtedly the radar display.

Radar displays. It is still unfortunately true that in spite of all the research which has been done by physicists and psychologists accidents at sea are still occurring with alarming frequency. Most collisions barely rate a mention in the press, but occasionally one does hit the headlines, as did that between the *Andrea Doria* and the *Stockholm.* Research has been concentrated on problems associated with the detection of the radar echo or pip, but no one seems to have been very much concerned with what happens afterwards. Even now, many skippers distrust the information which they get from their radar and prefer to rely on their own visual observation (there is some legal backing for this), so that we do not yet know the answer to the very important psychological problem as to why when he should be able to see that he is on a collision course with another vessel, a skipper will persist in maintaining his bearing and speed until the inevitable occurs. It must be added, however, that investigations of this kind are incredibly difficult and time-consuming; more often than not the accident would be due to a quirk of personality, which would make the implication of any findings extremely difficult.

The most commonly used form of radar display is that known as a plan position indicator (pip), which has a circular phosphor covered screen with a polar grid giving range and bearing. As the aerial rotates usually between 4 and 6 rpm the surface of the pip is swept by a bright line which leaves behind it pips that indicate the existence of objects producing echoes. If this was all that appeared on the pip there would be no very great problem, but unfortunately there is a good deal of background 'clutter' or 'noise' which can make the detection of a wanted echo quite difficult. Since the centre of the pip is the location of your own ship and this is moving relative to other objects around it, information has to be transferred onto a chart, the plots on which should show the relative movement of your own ship and those surrounding you. A similar situation exists for use with aircraft. In the military context, radar information is used in carriers for the remote control of their aircraft while in the civilian context it is used for the handling of aircraft at airports. In all these cases, it is vitally important that 'real' pips should be identified and kept under review.

The resources which have been ploughed into research have been very substantial. Johns Hopkins University had a special highly sophisticated laboratory built, the Systems Research Laboratory, in which they could simulate all kinds of layouts of equipment and compartments. The Naval Motion Study Unit was provided with facilities at radar establishments which enabled them to simulate whole radar operating systems and in particular to estimate the losses through the transmission of information via human links. It is true that much of the information thus obtained was of interest primarily to the military context but even so, much of this work has been very valuable for civil use.

It is impossible here to give details of even a small part of the research which has been undertaken (for a very clear summary, see Baker, 1962), but the research falls into two general areas. First, the effect of varying physical aspects of the display on the operatives' ability to see a pip; these include the brightness of the pip, the visibility threshold of the pip and the time that it persists (the decay time), the effect of the dimensions of the pip on its visibility, and the lumination which controls the contrast between the pip brightness and the brightness of the surroundings. Secondly, there are those things which are largely the function of the operator. These include such visual matters as adaptation of the eye, 'visual fatigue', studies of eye movement and search strategies, shape identification for coding purposes, and visual discrimination between adjacent pips which is an aspect of visual acuity.

Studies related to search strategy have included visibility thresholds, the effects of overall display size and viewing distance. There have been several studies of eye movement, and in particular of a curious ghost sweep which appears to be associated with an after image, and investigations into a number of different imposed strategies and the use of more than one operator. When a pip has been detected, it is necessary to report its position, and investigations in this area have included the accuracy of both range and bearing information which has been undertaken in relation to the design of the different polar grids which form the surface of the pip. In aircraft use a different display is often used for height but some experimental displays have also been produced which display the height on the outer edge of the pip; reporting on these has been compared with that for reporting from two separate displays. A number of aids to the reporting of range bearing, such as cursors, have been developed and these have been very fully investigated.

Once the operator has decided what the range and bearing of a pip is, this has got to be transmitted to somebody else – usually the man who is going to plot it. A first requirement is to ensure that the correct pip is being identified and pip coding has therefore been thoroughly investigated. The subjects of other investigations have been the effect of the quality of telephone lines, the effect of poor diction, of signal and noise ratio, and of the ability of the man at the receiving end actually to plot accurately the information which he has been given.

The radar operator as a monitor has also been thoroughly investigated. One of the earliest and certainly one of the most widely known experiments was that of Mackworth's (1950) clock test. This demonstrates a decrement in vigilance during which the alertness of the operator and his readiness to see a pip deteriorates. This work has developed into a much larger area of investigation which has nothing to do with radar operating and there seems to be some doubt as to whether it has any real practical consequences. In so far as it was applied in the radar situation, it was suggested that the length of watch should not exceed 30 mins. A major difficulty with research of the kind that was carried out on vigilance in the laboratory soon became apparent when attempts were made to apply the findings in a real-life situation. When trials were carried out aboard ship, for 'real', the laboratory results were not confirmed (Wallis & Samuel, 1961). It has been suggested that these differences lie primarily in the motivation of the subjects being investigated. Mackworth's subjects at Cambridge were a Naval Unit detailed to take part in experiments and as far as one could see they had no interest in the proceedings whatsoever. The same would seem to apply to the military personnel used by a number of investigators in the United States. But when you go aboard ship, or are flying in an aircraft, you are no longer in an artificial situation and your very life may depend on the detection and interpretation of a particular pip. Radio operators who are actually performing on board ship are therefore likely to be much more highly motivated than those who have merely been detailed as guinea pigs to sit in isolation for a boring one-hour session in a cubicle.

Digital micrometers

A micrometer is a device used by skilled engineers for measuring diameters. It is, in a sense, the badge of the skilled engineer and it is generally believed that extensive training in its use is required and that readings are made without error. However, there have been indications that this is not true. In one experiment skilled charge hands were found to have 8 per cent of the readings in error. The thimble of the standard micrometer was graduated round its edge from 0 to 25 thous with a scale mark at every thou; the barrel was graduated in units of 25 thous, the index for this scale being the edge of the thimble. Thus a reading of 83 thous will be made up of 3 × 25 read from the barrel plus 8 thous read from the thimble. It is very easy to make a mistake in this mental arithmetic – to misread the scale by 5 or 10, or the number of thous recorded on the barrel. It is not very surprising therefore that the initial experiment found that the errors were mainly of either 5, 25 or 100 thous.

It seems reasonable to assume that since digital presentation has been shown on a number of occasions to be superior to reading from a graduated scale, this would also apply to the reading of a micrometer.

Two different display designs were tested by Murrell and Kingston (1966) against the standard (X); one of these (Y) had the counting wheels concentric with the spindle with the numerals appearing as black on white through three small windows. The second digital micrometer (Z) had the display in the yoke with the counting wheels driven by a chain of gears; the numerals were white on black and appeared through a flat window. Twelve apprentices and eighty-nine journeymen drawn from eight firms were tested.

The twelve apprentices were divided into two groups of six; group A had received one and a half hours instruction on the standard micrometer while group B had been through a full course of training followed by two or three months use. The errors recorded for group A were 22·5, 1·7 and 4 per cent on the three micrometers X, Y and Z. For group B, who had had some experience, the corresponding errors were 7·5, 1·7 and 0·8 per cent. When the journeymen were tested they made 3·73 per cent errors on X, with 0·56 and 0·53 per cent errors on Y and Z respectively. In both cases virtually all the errors were misreadings in units of 5 or units of 10.

This experiment demonstrated clearly the advantage of using digital presentation instead of scalar on a long-established and essential engineering measuring instrument. There was no difference on the basis of accuracy but the time taken was significantly less with Z. What is interesting is that the performance of the apprentices on the digital micrometer after only one and a half hours training on a standard micrometer was substantially better than the accuracy of the journeymen on the standard micrometer.

Altimeters

Difficulty of reading the three-pointer altimeter was mentioned earlier in this chapter. The history of the development of this instrument has been reviewed by Schum *et al.* (1963) and by Rolfe (1969). The earliest altimeters were single-pointer instruments rotating anti-clockwise from 0 to 10,000 feet. When this altitude was exceeded after 1930 the single-pointer instrument was retained but was converted into a two-revolution instrument up to 20,000 feet on an outer scale and up to 40,000 feet on an inner scale, with a scale mark every 200 feet on the outer scale and every 500 feet on the inner scale. During World War Two 40,000 feet began to be exceeded and so the three-pointer altimeter became the most widely used instrument and was the one which was shown to be capable of being read seriously in error. Although Grether (1947) had clearly demonstrated this fact and had shown that instruments using digital presentation would substantially reduce the commission of errors nothing serious was done. So in April 1958 a BEA Viscount approaching Prestwick Airport ran into the ground almost immediately after the captain had reported that he was flying at 11,000 feet. Only ten months later in December 1958, a BOAC Britannia on a test flight ran into the ground shortly after the crew reported that they were flying at a height of 12,000 feet. These are just two examples of a number of accidents which can almost certainly be attributed to misreading this particular instrument. In both instances a modified version of the three pointer altimeter was in use in which the smaller hands had been accentuated. Of course the basic fault had not been corrected which was that the largest pointer which has the greatest attention-getting value tends to be read first as giving tens of thousands of feet whereas in fact it should be read last giving the hundreds of feet. In ordinary parlance you talk about twenty minutes past eight rather than eight twenty because you tend to read the minute hand first and the hour hand second.

In 1959 the Minister of Transport and Civil Aviation in the United Kingdom set up an altimeter committee to investigate the design of altimeter faces and mechanisms. One of the first experiments which was tried was in the form of an opinion survey in which a series of designs were circulated to air crew and ergonomists to obtain their views of a number of alternative designs primarily using digits with a hand retained for the first thousand feet. The second stage of the research was a laboratory comparison using a technique of subject-determined exposure time in which two prototype counter-pointer altimeters were compared, one having the counter inside the circular scale, the other having it outside; this caused the diameter of the scale to be somewhat smaller. These were compared with a three-pointer design which had an expanding ribbon and peripheral pointer to indicate the tens of thousands of feet. There was no difference between the two counter-pointer altimeters which had zero reading errors; however, 20 per cent of the readings on the multi-pointer display were in error by 100 feet or more, the majority of errors being between a thousand and ten thousand feet in magnitude. The mean reading time for the counter-pointer altimeter with the counter inside the display was 1·34 secs whereas that for the three-pointer design was 3·90 secs. The third phase of the research were simulator and flight trials. Prototype instruments were fitted in flight simulators and there emerged a clear preference for the instruments with the counter inside the display; there was a tendency for the instrument with a counter outside the display to be read as two instruments and to be wrongly associated with the vertical speed indicator.

Instruments were installed in two aircraft for flight trials and each was flown for 250 hours in each of the two aircraft. A total of 96 pilots assessed the two displays; no measurement was taken. The opinions of the pilots were collected and there was no overwhelming preference for either of the counter-pointer type instruments.

The outcome of all this research was that in 1965 the Ministry of Aviation recommended that for civil use a counter-pointer display should be adopted with the counter mounted inside the circular scale. This design of altimeter is now in wide use in civilian aircraft.

It is interesting to note that it took over twenty years and many hundreds of deaths before a poor display was replaced by a satisfactory one.

These are three examples of research into one specific type of display, in addition to all the basic research discussed in this chapter.

Psychologists have become involved in extensive research programmes on such things as air traffic control, process control in chemical plant, railway locomotive drivers' perception of signals and studies of different schedules of watch keeping. Other forms of special display which have been investigated from time to time are too numerous to be mentioned here; they include graphic displays in chemical plants which show the actual process and location of the valves and sensors for the dials, numicator displays which actually tell the operator in words what has happened or what has to be done, warnings and alarms both visual and auditory.

4
Man to machine

A man's method of communicating with his machine is for all practical purposes some form of muscular activity operating a control. Even if the device used is a voice key, muscles are still required to operate the larynx. On the whole, controls have been less studied by psychologists than by work physiologists, but psychologists have been involved in such things as the expected direction of movement, particularly in relation to a movement produced in a display (compatibility), the accuracy with which control movements can be carried out, and such miscellaneous matters as the different shapes of control which can be discriminated by touch. There are other important factors such as forces or torques which can be applied. The physical position in which controls should be placed in relation to the operator's body have largely been studied by the functional anatomist and the work physiologist. Specific design recommendations which have resulted from the research will not be repeated here – they will be found in the texts to which reference has already been made – but in this chapter we shall concentrate on various matters which have been the subject of extensive psychological investigation.

Before doing this it is necessary to present a list of the names of the different controls and these are set out in Table 4.1 with some of their characteristics.

The statement made at the outset that control is operated by muscular action may be slightly misleading since all muscular activity must be under neural control. Anyone who has read this text so far will have realized that an important factor in all control activity is feedback and that every possible effort

Table 4.1 Characteristics of controls (after Murrell, 1960a)

	Suitable for:			*Range of*	
Type of Control	*Speed*	*Accuracy*	*Force*	*Movement*	*Loads*
Cranks – small	1	3	4	1	up to 4·9Nm
– large	3	4	1	1	over 4·9Nm
Handwheels	3	1	2/3	2	up to 18·5Nm
Knobs	4	2	1	2	up to 1·6Nm
Levers –					
horizontal	1	3	3	3	up to 133N
vertical	1	2	Short 3	3	up to 133N
(to–from body)	1	2	Long 1		*
vertical (across body)	2	2	2	4	one hand 90N two hands 133N
Joysticks	1	2	3	3	up to 22N
Pedals	1	3	1	4	133–890N†
Push-buttons	1	4	4	4	9N
Rotary selector switches	1	1	4	4	up to 1·2Nm
Joystick selector switches	1	1	3	4	up to 133N

Code: 1 = best or most suitable
4 = worst or least suitable

* When operated by a standing person depends on body weight
† Depends on leg flexion. Ankle only, 90N.

should be made to make this immediate. In addition to any external forms of feedback, man has internal feedback known as kinaesthesis which is of great importance in work and is involved in every movement we make.

Kinaesthetic receptors are of three main types. The first has nerve endings in the muscle itself which can either be spirally wound around the muscle fibres or can have spray-like endings distributed over the muscle fibre. These end-organs are stimulated when the muscle stretches, but not when the muscle contracts. Secondly, receptors are located at the junctions

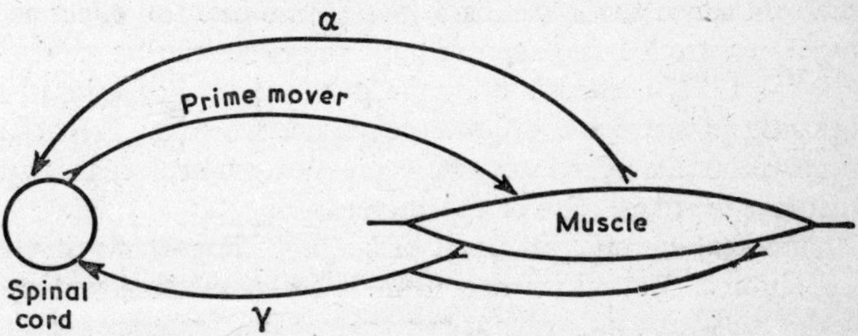

Fig. 4.1 *Feedback loop for control of muscle activity. α loop decreases rate of firing, γ loop increases it. The discrepancy between the two loops monitors the activity of the prime mover*

between muscle fibres and tendons and have spray-like nerve endings. They are stimulated either by a decrease or an increase in the tension at the junction, but it is not known whether they can distinguish between the two. These two receptors acting together will indicate whether the muscle is contracting or relaxing. The third type of receptor is not involved in movement; it is an encapsulated end-organ and gives information about deformation of the tissue in which they are embedded.

Kinaesthetic receptors form an important link in a feedback loop known as the gamma arc (Fig. 4.1). When a muscle receives a command to contract a certain amount, the kinaesthetic receptors feed back information of the actual extent of the movement to the anterior horn cells, which in turn influence the amount of stimulation being fed through to the prime mover, so that its stimulation is varied in order to achieve the desired result. This feedback loop is not instantaneous so that there is always some degree of 'hunting' which can be most clearly seen as tremor if the finger is held out at arm's length.

If a control movement is made 'blind' so that there is no feedback of the extent to which the hand has been moved, it is not too easy to monitor accurately either the distance or the direction. On the other hand, if a control is virtually isometric (without movement) and depends for its effect on pressure, then the kinaesthetic feedback loop comes into full effect and highly controlled and reproducable pressures can be exerted. Gibbs (1954) used this principle in the design of a joystick utilizing pressure and produced evidence which suggests that there is greater feedback with pressure control than with displacement control – of this more anon.

Another neurophysiological principle of importance in the operation of controls is antagonism. When a limb is moved, one set of muscles contracts but at the same time another set (the antagonists) will relax and regulate the extent of the movement by resisting the contraction of the prime mover. The precision with which movements can be made depends very much on the geometry of the various joints which will enable sets of muscles acting in antagonism to control each other's activity. Thus we find that precision of movement about the shoulder or elbow is very much greater than that about the hip or the knee. The same principle applies when two limbs are acting in antagonism to each other to produce greater precision of movement than when one limb acts alone; thus a handwheel (such as a car steering wheel) can be operated with much greater precision than can a lever which is pushed fore and aft and by the same token a rudder bar can be operated substantially more precisely than can a foot pedal. The rotary movement of the forearm in pronation and supernation, which is achieved by crossing the radius and the ulna (the bones in the forearm), also permits antagonistic forces to come into play to give precise movement in the operation of knobs; not as precise as a handwheel but substantially better than a lever. For the foot, a pedal operated by flexing and extending the ankle is operated more precisely than a pedal which is operated by flexing and extending the knee.

Yet another neurophysiological factor determining the precision of control operation is the enervation ratio. When a muscle contracts only a proportion of the muscle fibres which it contains will be involved in the contraction, unless a maximal force is being applied. When a movement is being made, its precision is controlled by the successive firing of the nerve endings attached to the muscle fibres. The most precise control is in the operation of the eye, where there is virtually one nerve ending to every muscle fibre. In other muscles, however, the number of muscle fibres is much greater and the larger the number of muscle fibres fired by each nerve ending the coarser will the movement be. In the arms, and in particular in the muscles controlling the fingers, the enervation ratio, that is, the number of muscle fibres fired by each nerve ending, is substantially less than in the leg and in the foot. So that purely from the point of view of muscular control, and quite apart from the geometry of the limbs, movements with the arms and hands are substantially more precise than movements made with the legs and feet. On the other hand, where sheer force is required, the substantially more powerful muscles in the lower limbs will give them a marked superiority over the remainder of the body.

Thus we see that purely from the anatomical and neurophysiological point of view the choice of type of control we use will depend upon the extent to which precision and/or force are required. A pressure type of control utilizing kinaesthetic feedback is probably the most precise; of the free moving controls, those utilizing antagonism – handwheels and knobs for the arms, a rudder bar for the feet – are most precise. When force is required leg muscles should be used although as an alternative, long levers which enable the whole of the body weight to be used can produce a very high torque.

Control dynamics

When man's basic output, force, is applied to controls with different characteristics, they will respond in ways which produce different outputs which can be given different mathematical representations. A simple control could be a spring-centred joystick which can be represented as in Fig. 4.2a.

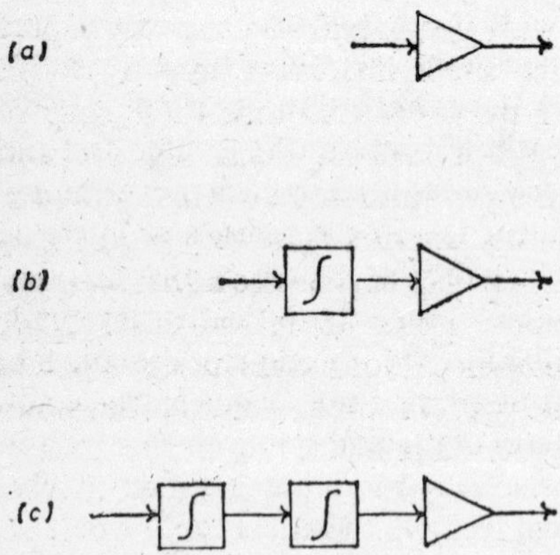

Fig. 4.2 *Dynamics of* (a) *a spring-load control,* (b) *a control with viscous friction, and* (c) *a control with inertia (from Murrell, 1965a)*

When force is applied the resulting displacement is directly proportional to its magnitude, but will vary with the extent of the spring loading. In other words, the joystick functions as an amplifier of adjustable gain or, to put it yet another way, the harder you push the greater will be the displacement. If instead of being spring-loaded the joystick has viscous damping it will move with an angular velocity, or rate, which is proportional to the force applied and the displacement is therefore proportional to the first time-integral of the force (Fig. 4.2b). In other words, the harder you push the faster will the control move. Now if force is applied to a joystick with high inertia, it will move with angular acceleration which is proportional to the applied force; the displacement is then proportional to the second time-integral of the force (Fig. 4.2c). In other words, the harder you push the greater will be the acceleration of the control. Thus depending on the dynamics of the control mechanism, the control output in terms of time/displacement is nought, one or two integrations of the force applied.

If instead of having controls with these characteristics a control is completely 'free', then the operator would find it difficult, even with visual feedback, to make the required control movements, especially when rate or acceleration are demanded. But when controls with these characteristics are used, the appropriate feedback comes into play and even control of acceleration, as for instance when starting up equipment at a controlled rate, becomes quite easy since the operator will rapidly learn the amount of pressure he has to apply in order to achieve the desired result. If he had to do it by estimating acceleration from a needle on a tachometer, which would be the only alternative with a 'free' control, he would be both unreliable and unpredictable.

Controlling by pressure

Research into the comparative merits of pressure and amplitude cues in the operation of controls are a nice example of the study of what are basically irrelevant questions. On the basis of evidence which he quoted, McCormick (1964) says: 'Certain studies have suggested that amplitude and movement is a feedback source superior to force'. The primary study on which he bases this view is that of Weiss (1953), who made his subjects operate a lever, in a blind positioning task, with varying loads and with varying amplitudes. He found that the distance travelled provided a better cue for positioning the lever accurately than did load. But how often does an operator actually use a control 'blind' in the sense in which this word is interpreted in this experiment? And how can you 'position' a control when the only cue is pressure. If we remember the discussion on control dynamics above (p. 78), you would see that what really should be compared is the ability of an operator to move a control at a particular rate or with a particular acceleration when compared with achieving the same objective by means of pressure. As has already been mentioned, one study which does make a comparison of this kind is that of Gibbs (1954), who compared a free-moving joystick (amplitude control) with a pressure (isometric) joystick for a compensatory tracking task. Both controls were so arranged that the greater the pressure or the displacement, the greater the speed with which a target would move across the face of a cathode ray tube (CRT). The subject's task was to maintain the moving target in the centre of the face of the tube by moving it back when it deviated as quickly and accurately as possible.

This type of system is known as velocity control since the speed of cursor movement is proportional to the extent of the control movement. An alternative system, which has been shown to be superior for most tracking or target acquisition tasks, is positional control in which the displacement of the controlled element is proportional to the displacement of the controlling lever. Burke and Gibbs (1965) compared a free-moving control with a pressure control in a positional control system with zero lag in which the displacement was proportional to the pressure applied (an isometric version of control type (*a*) in Fig. 4.2) in a pursuit tracking task. Pressure control was superior to free-moving control a $p < \cdot 001$. It was found that there was less error when the pressure control was used.

Although the blind positioning may be relatively irrelevant to the practical situation, ability to reproduce pressures will often be of importance. It is true that when you are applying the brake on a motor car (this, incidentally, is an isometric control when the actual braking takes place) you get immediate feedback both from the vestibular apparatus and visually as to the effect of the pressure you are applying, but if you are using an inertial control in order to bring a dynamo up to speed at a controlled rate of acceleration it is unlikely that you will get any meaningful feedback, and under these circumstances the ability to press the control with the right pressure becomes quite important. Jenkins (1947) investigated the ability of subjects to reproduce various forces using a lever, handwheel and a pedal. The results are very similar for all three types of control and suggest that provided the force required is greater than about 6 lbf (27N), the accuracy of reproducing standard pressures will be quite high. The upper limit is that which can be maintained without fatigue.

Tracking systems

A substantial amount of the research involving man as a controller in a closed-loop system has involved tracking, since this task is analogous to flying aeroplanes or driving motor cars. From the point of view of controls four main areas have been investigated: the control/display ratio, system lag, back-lash (continuation of display movement after control movement has stopped) and dead space (control movement which produces no effect).

The control/display ratio relates the amount of movement in a control to the effect which it produces in a display. The reciprocal of the C/D ratio is known as the display gain. When a small movement in a control produces a large movement on the display, the C/D ratio is low, and conversely when a large control movement produces only a small display movement, the C/D ratio is high. Most of the results are specific to the particular situation being investigated so that one can only state in general terms that as you move from a low C/D ratio to a high, the time taken to make a final adjustment will decrease, but the time taken to travel to the point at which this adjustment is made will increase. In a tracking task, however, the C/D ratio is not independent of other characteristics of the system.

In Chapter 2, process delays were referred to in the control of such things as chemical plants. These are analogous to system lags in a tracking system except that the latter are usually very short and occur between an operator's control action and the corresponding response of the system. Because the lag in a system is part of the function of the mechanical or electronic part of the system, it is of course possible, within reason, to redesign the system in order to change the lag. However, this reduction may involve engineering considerations and costs which are disproportionate to a potential gain and are therefore unacceptable. It will be remembered that in addition there will be human lag due to the time taken to react, and the two lags together may produce unacceptable system performance. There is evidence, however, that in some circumstances an operative can learn to anticipate what is going to happen and so reduce the effect of the combined lags.

Lag, together with back-lash and dead space, appear to interact not only with each other but also with the C/D ratio. All these aspects have been investigated thoroughly in combination for the United States Airforce in the early 1950s, mainly by Rockway and his collaborators (e.g. Rockway, 1954). One interesting finding from Rockway's work was that with an exponential time delay a low C/D ratio (for example in the region of 1:30) performance with lag would be an improvement over that with no lag. But in general the greater the C/D time lag the greater the degradation in performance with high C/D ratios. Rockway also demonstrated that there are similar effects with back-lash and dead space and the C/D ratio and that none of these three effects are entirely independent of the other. The importance of this work is largely restricted to the transport industry.

Other design details

Extensive studies have been carried out by the Psychology Branch of the Aero Medical Laboratory of the United States Airforce on knob shapes which can be distinguished by touch. There is some confusion over the term 'knob' in that it can be a knob which can be rotated, but can also be the bulbous termination of a lever. Both types have been investigated under the same name. In addition to coding by shape, coding by texture, by size, by colour and by location have been looked at. There are other aspects of controls which have been investigated by (mainly American) psychologists: these include the optimum diameter of cranks in relation to speed of cranking; the location and direction of movement of cranks in relation to two-dimensional tracking; the amount of movement in a control when making settings by 'bracketing'; the nature of the cues required (for example with the use of detents – stops which retain a rotary control in a series of discrete positions) for discrete switching and blind location of controls in a reach envelope around the human body. Design recommendations based on this work will be found in appropriate Cook books!

Compatibility

In our discussion above on time lags in a system we referred specifically to the importance of determining the optimum relationship between the movement in the control and the display. In the various studies of control operation in relation to display effect it quickly became evident that the physical relationship between the two was important; that there were certain control movements and display effects which were expected by the majority of the population, and that it is important to know what these expectations are and what is their strength. We do not need to be told when we get into an unfamiliar motor car that we will have to turn the wheel clockwise to make the car turn to the right; that we will have to turn the control knob of a radio set clockwise to turn it on or, if we have to tune the set, that a clockwise movement of the control knob will produce a left-to-right movement on the tuning dial. There are two types of control movement, one of which will produce an expected effect and the other an expected movement in its associated display; those which are expected by the majority of the population are often called population stereotypes and control/display movements which conform to these stereotypes are said to be compatible.

Population stereotypes

A question which is regularly raised in discussion of population stereotypes is whether these can be considered as being learned patterns of behaviour or whether some are inherent. In some instances the answer is quite obvious; in the UK the usual direction of the movement of an electric light switch is down for 'on', but in the United States it is down for 'off'. In other instances a stereotype may be inferred from the anatomy of the limb used for making the control movement, as when supinating the right hand in a *clockwise direction*, a movement which appears to be preferred to the *anti-clockwise movement* of pronation. Just watch anybody approaching a door – they will almost invariably turn the door knob clockwise in order to open it and this is just as certainly a consequence of the anatomy of the forearm and wrist. Sometimes it is difficult to conceive any other relationship being used. It is surely unthinkable for any person, however primitive or inexperienced, to expect the steering wheel of a car to turn anti-clockwise to turn the car to the right.

The question of whether the stereotype is natural or learned may at first sight seem unimportant, but there may be a conflict between a learned pattern and one which is natural and this can lead to difficulty or even disaster. It is well known that movement of steam valves or of ordinary domestic taps is anti-clockwise to increase the flow. But if clockwise-to-increase is a learned stereotype, the anti-clockwise-to-increase movement can also be learned and no harm is likely to come in an emergency whichever relationship is used; but if the former is natural and the latter is learned, in an emergency someone may revert to the clockwise-to-increase stereotype. There is evidence that this will actually occur.

There appears to be relatively little evidence as to whether stereotypes are learned or natural since most experimental investigation on population stereotypes has been directed towards discovering what and how strong these stereotypes are. One investigator who has explored this matter extensively is Shephard at Toronto (e.g. Cook and Shephard, 1958), who has tested child subjects down to five years of age with three different control/display relationships on the Toronto Complex Coordinator, on which the movement of a vertical joystick controlled a circular marker which had to be moved to encircle a target; successful holding of the target for a short period of time was described as a 'match'. A number of other scores were possible and one of these was the number of occasions on which the controls were moved in the wrong direction when starting towards a fresh target. The compatible arrangement was away for up and left/right for left/right. Now the latter movement may perhaps have entered into the experience of very young children, but it seemed highly unlikely that any child would have experienced the former movement. Alternative arrangements were tried with the joystick turned clockwise through 90° to the 3 o'clock position and through 180° for the 6 o'clock position, the latter arrangement being directly opposite to the compatible arrangement.

Three groups of children of both sexes were tested, their ages being five, ten and twenty years. It was found that the boys were on the whole better than the girls in the number of matches made in a given time, but this difference was less marked in the five-year-olds with the compatible arrangement and, considering the five-year-olds only, approximately four times more matches were made on the compatible arrangement than on the incompatible, 6 o'clock arrangement. There was a substantial improvement in performance with practice at this age, but the relative superiority of the compatible arrangement remained comparatively unchanged. Initial movement in the wrong direction, which can be a very good index of the strength of compatibility, was made in about four-and-a-half out of ten matches with a compatible arrangement compared with six-and-a-half out of ten with the incompatible arrangement. With increasing age, errors of this type decreased substantially to only one-and-a-half in ten at age twenty, which may perhaps indicate that greater experience in the manipulation of controls may make it easier to learn an incompatible arrangement.

Evidence from this one piece of research may perhaps be taken as indicating that even in young children of the age of five years who, it would seem, are unlikely to have experienced any particular control movement, there is a stronger tendency to operate the control in the compatible direction rather than in the incompatible direction. It would seem that this piece of research remains unique and in the absence of other experimental evidence one should look elsewhere for information. Accidents are probably one of the best sources from which inferences can be made, provided they are properly reported.

One such accident is quoted by Murrell (1965a). A lubricating oil cooler in a ship was controlled by a valve which turned, as is usual, anti-clockwise to open; above this there was an indicator with three positions (Fig. 4.3). As the control valve was turned anti-clockwise the indicator moved clockwise (from the left to right) from the 'off' position to the 'on' position; beyond the on position was the third position, 'bypass'. The valve was set in the 'on' position while work was being carried out on the cooler, when an emergency suddenly arose and someone yelled to a sailor standing nearby to bypass the filter. Seeing that he had to move the pointer from the centre position to the right, the sailor turned the valve clockwise thereby closing it, with the result that the cooler was seriously damaged. This and similar accidents suggest that in spite of the learned habit of turning the handwheel anti-clockwise to open, in an emergency the sailor reverted to the 'natural' stereotype of clockwise to open even though this movement was contrary to a pattern which he would have been expected to have learned. You need not be told what happened to the unfortunate sailor; the person who really should have been 'carpeted' was

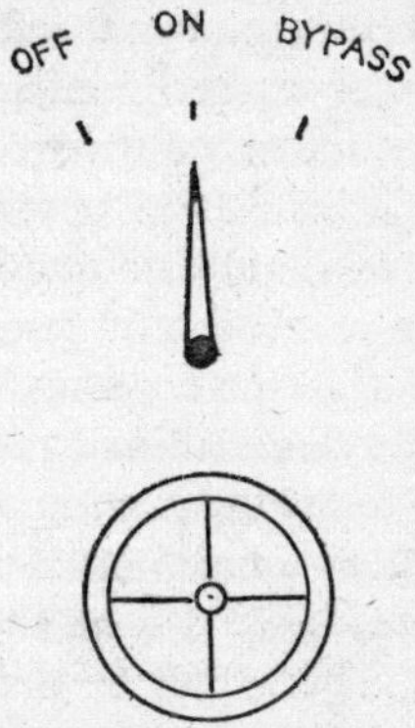

Fig. 4.3 *Control for a lubricating oil cooler (from Murrell, 1965a)*

the designer, for an accident of this kind was completely predictable.

To carry this matter a stage further, let us consider the hypothetical case of two matched naive men who are taught to drive cars; one (N) has a car with normal steering and the other (R) a car with reversed steering in which the wheel is turned anti-clockwise to make the car turn to the right. Both will learn to drive but R will take much longer to reach the acceptable standard of proficiency than will N. Now let us assume that both men have nagging wives and that both take these wives for a Sunday afternoon drive; the 'back seat' tactics will have relatively little effect on N but the driving performance of R may become quite bad. A child runs across the road: N takes a correct avoiding action but R reverts to the population stereotype, turns the wheel the wrong way and kills the child. Although this example is hypothetical, the effects described have been found in a number of laboratory experiments. For example, Taylor and Garvey (1959) trained subjects on a tracking task with positional and velocity controls, movement with the latter being incompatible over part of the task in that the control must be moved, say, to the right to slow down the marker which was still moving to the left. Initial performance with velocity control was markedly inferior to that with positional control but this difference disappeared after seventeen sessions. It reappeared when 'stress' of various kinds was added to the main task. Simon (1951) also found that a long period of training would produce a satisfactory performance with a non-preferred relationship, but under stress an operation would revert to the preferred relationship.

From this work the following general conclusions have been drawn: (i) Learning time for the operation of equipment on which controls are compatible will be shorter than if controls are incompatible. (ii) There will be a much greater risk of accident on equipment with incompatible controls, since in an emergency the operator will tend to revert to the natural direction of movement, however well the incompatible movement has been learned. (iii) Performance on equipment with incompatible control movements will deteriorate if any form of stress, even one as simple as dealing with mental arithmetic, is put upon the operator. It can be expected, therefore, that the output of machines on which controls are incompatible will be inferior to that of machines with compatible controls. (iv) All controls on a given piece of equipment should be consistent in the effects which they produce and all controls used for a given purpose, for instance starting up or closing down a piece of equipment, should all move in the same direction; as far as possible, the same control of the movement should be used on all pieces of equipment, especially when an operator is likely to move from one piece of equipment to another. If complete compatibility cannot be achieved, it is better to have all the controls incompatible than to have some compatible and some incompatible. (v) Some stereotypes are likely to be much stronger than others. So, if emergency operation is at all probable, the control movements which have strong stereotypes should be selected in preference to those which do not.

Relatively little research has been carried out on the expected *effect* of control movements, since most has been concentrated on the situation where a display such as a dial is involved. Nevertheless, it does appear that there is a fairly strong stereotype of clockwise for 'on' with a rotary control, up or away for 'increase', and left–right for left to right and vice versa. Agricultural machinery, particularly the tractor, is a good example of inconsistency in control operation. On some makes the throttle control goes forward to increase speed which appears to conform to the stereotype, whereas in others it is moved backwards. But the same tractor which has forward-for-increase on the throttle may have a backward-moving lever to lift the hydraulics, and manufacturers seem totally unable to agree on what movement is correct for this function since on some makes of tractor the control lever is pushed forward to raise the hydraulics. Anyone switching from one make of tractor to another, as is not uncommon on British farms, could easily have a mishap, partly because of the inconsistency of the movements between one machine and another and partly because a control movement may not follow a natural stereotype.

Actually, the tractor example illustrates the borderline between those stereotypes which have been extensively investigated and those which have not. The speed with which the tractor moves, as controlled by the throttle lever, is an effect which could hardly be described as a display. However, the raising and lowering of the hydraulics of a front-end loader is a display with a moving element which corresponds to the movement of a control lever. In most modern tractors, the front-end loader is operated by a velocity control with the speed of rising or falling depending on the displacement of the control from a central position, whereas the rear-end hydraulics are operated by the same control in a positional mode in which the amount by which the three-point linkage is raised depends on the position of the control lever. Although compatibility between display and control movements has been extensively investigated, the display has almost invariably been some kind of dial, circular or straight, and conclusions about other situations such as the tractor, the bed of a planing machine or the platen of a press have been by extrapolation or analogy, it being assumed, almost certainly correctly, that if a forward movement of a control is associated with an upward movement on a dial, then it will also be associated with an upward movement of the bucket on a front-end loader.

Display-control relationships

It is not known when it was first realized that the relationship between display and control might play an important part in efficiency of operating equipment. There is no doubt that some fairly obvious relationships were established by trial and error – at least they seemed so obvious that no one would question their validity: the steering wheel of a motor car is a case in point. The first really systematic investigations started during World War Two, in particular in respect to flight instruments.

One such instrument which came under early scrutiny because of the number of reported misreadings (Fitts & Jones, 1947) was the artificial horizon. It would appear that when the artificial horizon was first designed, the instrument was thought of as being, as it were, a window to the outside world. The aircraft was therefore kept level and the horizon tilted which is the same presentation as the pilot would see if he were looking out and not using his instrument. However, it soon became apparent that the information given by the instrument when the pilot was flying on instruments alone might cause confusion. The aircraft shown in Fig. 4.4 is banking to the left; when it has to be brought back onto an even keel the horizon should be turned anti-clockwise whereas the control movement involved should turn the aircraft clockwise. The control and the display appear to be incompatible, and this led Loucks (1945) to carry out an experiment on 555 subjects in a Link Trainer using four alternatives to the standard type. He found that performance on an instrument in which the horizon remained horizontal and the aircraft tilted was significantly superior to performance on other designs. He records also that this particular design was consistently preferred by a very large majority

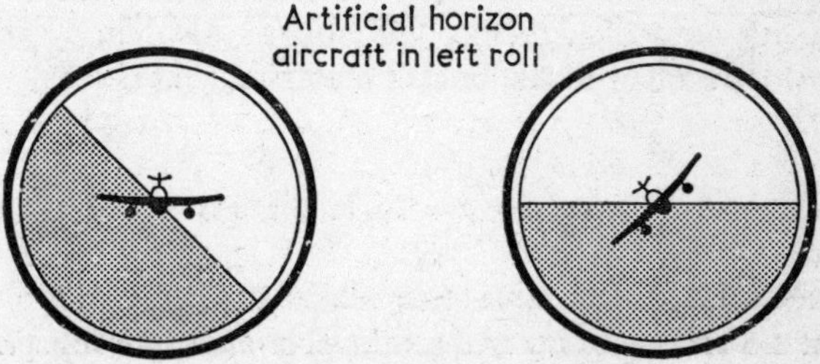

Fig. 4.4 *Comparison of incompatible* (left) *and compatible presentation of aircraft attitude*

of the subjects, some of whom actually turned their aircraft 'upside down' with incompatible arrangements.

A finding of this type can present a very difficult administrative decision. Most of the pilots flying at the time were accustomed to the old presentation. If a new presentation was adopted, would the learned pattern be too strong or would a change to a display which is more compatible be rapidly learned and would there be no tendency to revert in an emergency to the learned pattern. Unfortunately, there is very little experimental evidence on this point and it seems the only way in which it can be resolved in the practical situation is so radically to alter the display that the pilots cannot be confused. In the event, this is what ultimately did happen.

Apart from these specialized studies, a pioneering investigation was carried out by Warwick (1947a), who used five panels or boxes, as shown in Fig. 4.5. The display consisted in each case of a row of small lights and the control was a knob which, when turned, caused a light signal to move by the lighting of the lamps in sequence. Fifty subjects were required to centre the light from various positions to the left or right using the control knob. This result would be achieved whichever way the knob was turned and the experimenter recorded whether the initial movement was made in a clockwise or anti-clockwise

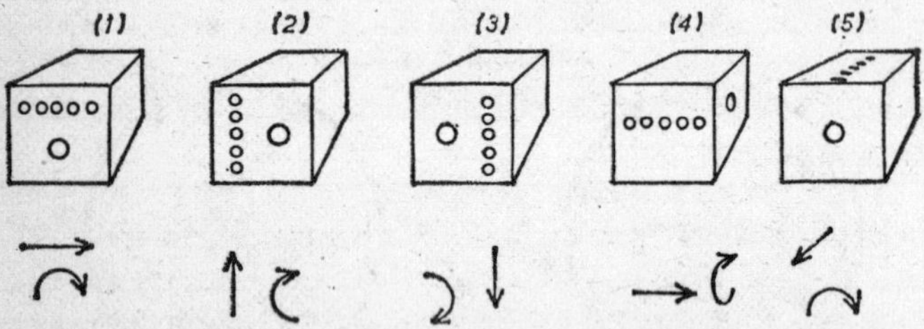

Fig. 4.5 *Arrangements of control and display with strongest preferred relationship* (*after Warwick, 1947a*)

direction. With arrangements (1), (2) and (3), in which the control knob is in the same plane as the display, between 70 and 95 per cent of the subjects moved the control clockwise to produce a left/right movement in arrangement (1), an up movement in arrangement (2) and a down movement in arrangement (3). When the control and the display were in different planes, as in arrangements (4) and (5), the preferences shown by the subjects were nothing like so clear cut and there appeared to be no general expectation of a particular direction of movement relationship.

Warwick therefore conducted a second experiment using a further forty-two subjects in which the relationship between the control movement and the movement of the light was fixed. The experiment was carried out under conditions which would enable him to determine which arrangement would produce the greatest speed with the smallest number of errors. For arrangement (4), this condition was satisfied when a clockwise movement of the knob produced a right to left movement of the lights, and for arrangement (5) when the clockwise movement of the control would move the light towards the subject. Warwick pointed out that the differences between these arrangements and alternative arrangements were quite slight and would probably have very little practical effect on efficiency. As a consequence, he was able to lay down a general principle that the control and the display should be in the same plane wherever possible and that the greatest degree of compatibility is achieved with a rotary control when that part of the control nearest the index of a display moves in the same direction as does the index.

Since Warwick's early work, a great deal of research has been carried out by a number of investigators, to the extent that in 1953 over one hundred references were quoted in a bibliography produced by Andreas; a more recent review has been made by Loveless (1962). The major part of this work has been concentrated on rotary controls associated with an index on either circular or straight scales and the use of joysticks and levers for different types of tracking; but the use of levers for making settings on graduated scales has received relatively little attention.

All this subsequent work seems to confirm the original findings of Warwick but with some reservations. First, when there is no continuity between the plane of movement of the control and the direction of movement of the display there may appear to be predictable expectations, but these may not be the same for all operators and, in some instances, the same operator may have different expectations on different occasions. Secondly, there may be a conflict between the expected direction of movement of a control without a display and the same control of movement when it is associated with a display. For instance, a lever placed beside the body may be expected to be pushed down to increase speed, but if the same lever is placed in front of the body and is associated with a vertical scale which moves upward to show an increase in speed then the expected movement of the lever may be upward. Thirdly, expectations which might be predicted on theoretical grounds may not be found in practice due to some peculiarities about the design of a particular machine or because of habits previously acquired by a particular population. Fourthly, an expectation may depend on whether the movement produced is left-to-right or rotary. The problem is that it is not easy to say at what degree of curvature the direction of movement of a pointer ceases to become rotary and becomes translatory. On a circular scale a downward movement of the pointer on the right-hand side will indicate an increase, but if only the right-hand sector of the circle is used a downward movement of the pointer would almost certainly be expected to produce a decrease.

This point was extensively studied by Loveless (1956) who found that subjects tend to respond to the rotary sense of the movement, that a clockwise control movement is expected to move a pointer to the right and that there is a tendency for the performance at the top of a scale to be inherently superior to that at the bottom. On quadrant scales there is a general tendency for performance to improve as the pointer is moved from the bottom to the top of the scale; but there is an expectation that a clockwise control movement will result in a clockwise pointer movement and that a clockwise control movement will move the pointer to the right or upwards according to the target used. With vertical and horizontal straight scales the best form was found to be a horizontal scale with a clockwise control movement sending the pointer to the right. Additional research has borne out that a circular scale with the pointer at the top and a circular scale with the pointer at the left, all associated with a clockwise movement, give the best performance.

From all this research on direction of motion stereotypes it is possible to arrive at a number of fairly well-defined expectations which should be used in design whenever possible, and these are shown in Fig. 4.6.

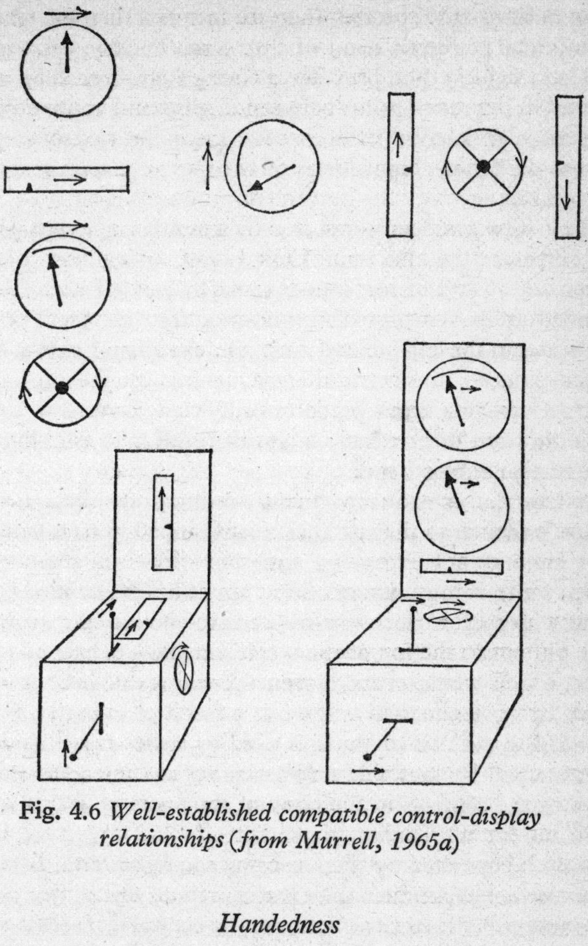

Fig. 4.6 *Well-established compatible control-display relationships (from Murrell, 1965a)*

Handedness

In most of the experiments quoted, the right hand seems to have been used, though whether the right hand was the preferred hand of all subjects tested is not always made clear. The research provides relatively little evidence of whether the same relationships are expected when the left hand is used, whether this is the preferred hand or not. Such evidence as there is tends to suggest that, provided a strong stereotype is involved, the expected relationship between display and control will be the same whichever hand is used since the relationship involves the physical movement of two parts one with respect to the other.

This view has been supported by a number of investigators; for instance Suddon and Link (1959) using the Toronto Complex Co-ordinator tested 15 right-handed and 15 left-handed subjects using their non-preferred hands. The performance of the left-handed subjects was superior to that of the right-handed subjects and this may be because the left-handed people have to live in a predominantly right-handed world and therefore can more readily adapt themselves to working with the non-preferred hand.

When the control movement produces an effect there is some evidence to suggest that a left-handed person using his left hand would expect an anti-clockwise-to-increase movement for the same anatomical reasons as a right-handed person would expect a clockwise-to-increase movement. However, experimental evidence does not confirm that left-handed people have this expectation and it seems probable that here one has a very strong learned stereotype as a result of living in a right-handed world. Which hand is used by either right-handed or left-handed persons will very much depend upon the orientation of the controls to the body of the operator and it will be difficult for a designer to predict whether a control would always be operated by the left or by the right hand. It is clear that further experimentation is required on this subject, but in the absence of further evidence to the contrary, it seems to be a fairly good rule that designers should use the same expected relationship for both hands.

Earlier in this chapter the importance was emphasized of the same C/D relationships being used throughout a particular piece of equipment and between different pieces of equipment. This may not always be the case since controls on most electrical instruments turn clockwise to increase while valves on equipment controlling liquids or gases turn anti-clockwise to open or increase. But the instruments which will be used with all these controls will invariably move clockwise to increase if they are circular or eccentric. If there are electrical and pressure controls mounted on the same panel there will thus be a compatible clockwise-clockwise arrangement with one but not with the other. The effect of this kind of arrangement has been studied by Warwick (1947b), who found that if compatible-incompatible C/D relationships were alternated, the subjects operated the controls far less efficiently than when both C/D arrangements were compatible. From this it follows that it can be more important to get a consistent relationship between a number of displays and their relevant controls, than to accept a compromise where some but not all of the relationships are expected. It further follows that if for some reason it is quite impossible to get an expected relationship between one or more of the displays and their associated controls, it would be far better to have all displays and controls incompatible than to have an inconsistent relationship between them.

In order to achieve a compatible arrangement when mixed controls are mounted in the same panel, valves which produce an increase in pressure would have to be fitted with left-handed threads if they are not remotely operated. It has been objected that the fitting of left-handed threads to valves would conflict with the learned pattern of an engineer who has been brought up in the operation of steam or hydraulic equipment. But will an engineer, seeing a knob protruding from the front of a panel with a gauge associated with it, still think of it as a valve or will he think of it simply as a control which is associated with the pressure gauge above it? This point was tested in an experiment in which a number of engineer officers and engine room petty officers and ratings were asked to indicate which way they would turn valves in the form of hand wheels protruding from the front of a panel in order to cause a given pointer movement (Murrell & McCarthy, 1951). The results showed that the large majority had a 'follow the pointer preference' as proposed by Warwick, although it was made absolutely clear to them that they would be operating valves. It would appear to follow that when a valve is mounted in a panel so that the pipework cannot be seen, the hand wheel controlling it becomes just another control with the normal C/D relationship. Left-handed threads should therefore be used to achieve this result.

Another aspect of consistency in the range of controls is when they must be used to close down a plant in an emergency. There seems to be very little doubt that all controls used for this purpose should turn in the same direction, whether they increase or decrease the function being controlled. If it is only a matter of seconds before a plant blows up, it is most unlikely that the operator will be able to remember that, for instance, the first three controls should be turned anti-clockwise, the fourth clockwise, the fifth and sixth anti-clockwise, even though under normal working conditions he may be able to remember these movements quite easily. If as a result of this, incompatible control display relationships have to be used for normal operation, this can be overcome by either using scales which are graduated anti-clockwise (there is some evidence to show that this is an entirely satisfactory arrangement) or a moving scale instrument may be used with a fixed pointer which will have the same effect. But whatever arrangement is adopted, it is very important that the probable human reactions to an emergency situation should not be forgotten or misinterpreted.

In this chapter we have looked at a variety of design points which have been investigated by psychologists at one time or another. Some of them may be quite trivial, but others are of great importance if efficiency and avoidance of accidents is to be maintained. It is entirely proper that psychologists should be involved in this kind of investigation since the questions being asked are all essentially behavioural ones and the techniques which need to be used are those in which psychologists are particularly expert. There are still many important points which have not been investigated and it is indeed unfortunate that the behaviour of rats or monkeys seems to hold much greater attraction than the behaviour of men in their working environment.

5
The interface

The interface is the imaginary plane which separates man from machine and across which information flows from the machine to the man and commands from the man to the machine. Although this is a conceptually correct statement, in practice the interface is a space which contains the means of communication, the displays and the controls. A great deal has been written about this interface area (e.g. Kelly, 1968) and there has been much research on some general principles but more on specific situations. The whole process of determining the optimum interface has come to be called *systems design* and the main purpose of people now working in this field is to develop a taxonomy rather than to discover a series of unrelated design criteria which may be of extremely limited application.

The purpose of the interface is to match as far as possible two widely differing components and this can be done on what may be called the physical and the psycho-operational levels.

Physical needs

Information on the physical requirements of a man–machine match across an interface have come mainly from the functional anatomist, the physical anthropologist and the work physiologist. In order to do his job a man must be able to see or hear what is going on, he must have space to move about in or to sit down when necessary and he must be able to reach and operate the controls without strain or physical injury. Unfortunately, instead of 'thinking through' each new design from first principles, too often designers seem to have 'the Nelson complex' and continue to perpetuate design faults and perpetrate abuses of the human body. The story of how the 'Nelson complex' arose is too long to tell here; suffice it to say that, when the idea that a coxwain of a motor torpedo boat should sit down to drive the boat, one senior officer said that he didn't like the idea 'because after all Nelson stood up to steer a boat', or in other words 'what was good enough for grandpa is good enough for me'. Thus we find machine tools which seem to have been designed to be operated by gorillas (for an example see p. 45 of E1), typewriters which seem to be designed to be used by a praying mantis, or fork-lift trucks which seem to operate on the assumption that the driver has an additional eye in the back of his head. All this is totally unnecessary because all the data which are required for the proper arrangement of the work space have been available for many years in a variety of publications.

The Nelson complex is clearly seen in the design of railway locomotives. Somewhere back in the 1880s Siemens developed the first electric tram car; speed was controlled by a system of rotary contactors operated by a crank on top. In due course the handle of this crank was developed into a 'dead man's handle'. Now the driver of a tram car normally stood up and with his upper arm vertical and forearm horizontal the weight of his arm could press down the dead man's handle without any great strain. Then the London, Brighton and South Coast Railway electrified some of its track and they too had contactors still with the handle on top with the dead man's function. But this was now operated at arm's length by a seated driver because the large contactor box made it impossible to get close enough to the handle to do otherwise. And so it has remained ever since:

the driving controls are still out in front of the driver, and the dead man's handle still has to be pressed down, an activity which can become very tiring. On one occasion, a driver was observed to place one end of a piece of wood under a suitable projection so that it lay across the dead man's handle; he looped a piece of string round the other end, into which he put his foot. He was then able to sit back in comfort. Why, one may ask, is the rotary movement about the shoulder not used with the main controls alongside the driver instead of in front of him? With this arrangement the weight of the forearm could again be used to hold down the dead man's handle. This is a logical arrangement well worth exploring, but perhaps it is too great a departure from convention. After all, it is not the way Siemens designed his tram car in 1885.

When the typewriter was first invented by Alexander Graham Bell the mechanical linkages required a keyboard which would be operated by both hands, palm down in a horizontal position. Now if the hand is relaxed it will normally turn so that the palm faces inward, and it takes continual muscle activity to keep the palm down. It always seems to be assumed that even when there are no mechanical connections, keyboards should be horizontal, and this is certainly true of the IBM card punching machines, one of which was studied by Yllö (1962). Yllö was the works doctor at the Volvo Skovde Works at Göteborg, and was interested in ergonomics. He became concerned at the amount of hyper-functional myalgia (inflammation of the forearm) among the operators of the card punching machines. He found that the keyboard was too high in relation to the working arm and was in a position which caused the forearm to be twisted with the wrist in an unnatural position. He used electromyography to establish the best position, which was that the keyboard be located close to the operator's right knee and tilted at an angle of about 45°. When these changes had been made the number of girls reporting inflammation of their forearms dropped dramatically.

These are but two examples of situations which are legion.

Most working situations can be faulted in one way or another and are only acceptable because of the enormous adaptability of the human body; but this is not by any means always without cost. Perhaps, taking a leaf out of the book of the Lord High Executioner and making the punishment fit the crime, designers should be made to work for periods on the equipment for which they are responsible.

The psycho-organizational function

Important though the physical arrangement is, of even more importance is the manner in which the interface permits the operator to receive and process information and organize his responses. To ensure that this is being done optimally it is necessary to ask and to answer a number of questions. These appear in various forms in the literature but can be summarized as follows.

What exactly will the man–machine unit be expected to do? Will it function on its own or will it be integrated with a series of other units? Will its input depend on the successful operation of a preceding unit and/or will its output determine the operation of a succeeding unit?

What functions can the machine do best and what rôle is the operator expected to play. Will optimum use be made of the inherent capacities of both the machine and the man and will either be called upon to undertake functions which they cannot do very well? If this is so, can these functions be transferred economically to the other partner?

What information does the operator need in order to do his task? In what form must he receive this information and how can it best be displayed? Should it be visual, auditory or tactual? If the information is to be given by a display, which type will give the information most quickly and with the minimum of ambiguity? Will the mental demands of the task be such as to overload the operator under extreme conditions and if so what steps can be taken to reduce the load?

What input does the machine require in order to function efficiently and can this input be provided by the operator? Which is the best type of control to produce the required input and what should its dynamics be? What is the time factor involved in successful operation of the equipment and can the operator meet this requirement? If executive action has to be taken which does not directly involve the machine, have adequate communication channels been provided? What are the economic consequences of various possible alternatives? Can the desired hardware be manufactured economically – if it all? And is it likely to be as reliable as possible alternatives?

Much of what has been said in earlier parts of this book will enable these questions to be answered. We know already those functions which are best assigned to a machine and those which are best left to the operator, but before this knowledge can be applied a very detailed task analysis will be required.

This task analysis should not only encompass the situation when everything is going smoothly, but should also, by the use perhaps of a little crystal ball gazing, tease out all the situations in which the operator may have to act quickly and decisively. An example of this in the operation of a chemical plant was given earlier (p. 33). Everything may be ticking along quietly with the operator working well within his capacity, but if something starts to go adrift then the situation may change suddenly and dramatically with the demands on the operator increasing so much that he may go into a state of overload, that is, he cannot react fast enough to cope with all the demands which are being made upon him. This concept of overload has been examined by Davis (1958) in a study of transport accidents in which he suggests that overload shedding will often take the form of making assumptions based on previous experience rather than on the ongoing situation. If the assumptions are correct, the emergency will pass and no one will be any the wiser. If, on the other hand, the assumptions are wrong, then at the best some small mishap may occur and at the worst there may be a devastating accident such as the Lewisham train collision in December 1957. These assumptions Davis calls 'false hypotheses' and he points out that under certain circumstances an individual uses a false hypothesis as a kind of defence, clinging to it despite clear evidence that it is false. A false hypothesis may also cause someone to concentrate on coping with a situation which he believes exists, while completely failing to see other evidence which shows that his hypothesis is wrong.

It may not be possible to predict all eventualities, but many situations can be foreseen and it is important that the interface should not be designed so that it misleads the operator. For instance, in a chemical plant certain compressors misfunctioned in the small hours of the morning setting off a loud alarm with lights flashing on the relevant part of the panel. The operator, who may perhaps have been dozing, suddenly leapt into action and tried to get the compressors back on line, completely forgetting that the first thing he ought to have done was to shut down another part of the plant. Since this part of the plant was still functioning normally there was no alarm to call the attention of the operator to the first action he ought to take. The result was a 24 hours shut-down instead of only perhaps 20 minutes. Now most of the things that could go wrong with this plant were known and were set out in an operating manual, but no one apparently thought of devising a comparatively simple scheme of telling the operator just what he should do by means of an annunciator panel (a panel giving information or instructions by means of words). Far from telling him what he ought to do, the whole design of the interface persuaded him to do the thing he ought not to do.

Where an operator has to make decisions, task analysis should tease out exactly what information the operator is going to need. In an early chapter, reference was made to the need for rate of rate of change to be displayed, but too often it is lower-order information which is given and the task of the operator therefore becomes that much more difficult.

When the task of the machine and the man have been analysed and it appears that the operator may either regularly or on occasion have to relate a series of information sources, it is necessary to decide how these should be arranged. If the new task is a variant of something which has gone before it is often possible to do eye movement studies of someone doing the task for real. One of the earliest reported uses of this technique came from the Psychology Branch of the Aero Space Medical Laboratory USAF, who used a camera in a flying laboratory in order to establish link values (the number of times the eyes move from one instrument to another) between the different aircraft instruments together with the length of eye fixations, the number of fixations and the proportion of time each instrument was looked at during a climbing manoeuvre with constant heading (Jones *et al.*, 1949). Patterning according to link values when undertaking a specific task may or may not conflict with scanning patterns under normal watch-keeping circumstances. Murrell (1960b) reported that there was a significant difference between the speed with which deviations were noticed in the top left-hand corner and the bottom right-hand corner of a panel as compared with the other two corners and this was later confirmed in an unpublished eye movement study which showed that the scanning pattern tended to be the same as that used when reading a book and that there was a tendency to dwell more at the top left and bottom right hand corners than elsewhere on the panel. The camera method of studying eye movement is somewhat difficult and very time consuming and a number of other techniques have been developed. The most successful seems to be electro-oculography, which was used for job design by Shackel (1960) and has now become a standard, commercially available technique.

In all the foregoing it has been assumed that the operator is comparatively static and that the space occupied by the interface is quite small. But this will not always be so; some control rooms are very large and although the operator may have a static console containing some of the major controls, and the dials may be large enough for him to be able to read the important ones from his fixed position, he may often move around from unit to unit making minor adjustments. In this situation, also, link studies and a prediction of likely upsets may save an operator from having to shuttle backwards and forwards across a control room (possibly so far as the designer is concerned in zero time) when he is trying to put things to rights. It is true that good functional arrangements in a large control room will be the antithesis of the beautiful, orderly, symmetrical arrangement which is so beloved by designers in general and management in particular. It is not unknown for a control room to be designed as a show piece with roped-off areas for goofers, illuminated alcoves resplendent with fresh flowers and views of the outside world in tempera. But they can be terrible places to work in and can lead to inefficient plant operation.

There are many situations where the man–machine interface does not involve any dials and perhaps only one or two controls. Into this category would come the machine-minder who looks after a group of automatic machines, ensuring that the tools are sharp and that they are kept fed with raw material; or the operator on draw-twisting of nylon or terylene where the criterion of success is keeping the maximum number of frames running and producing bobbins without flaws. In these situations the basic design philosophy is just the same as it is when designing a control panel. It is necessary to decide the extent to which it is economic to get a machine to look after itself, how it will tell the operator that it needs attention and how that operator can give it the attention it requires and get it back on line with all possible speed.

Enough has been said for it to be clear that establishing the interface is not a case of following a series of specific design recommendations; there are general design recommendations, of course, but they will be relevant only to a comparatively restricted range of situations. In most instances an interface will be a one-off job (this does not mean there may not be a great many interfaces of the same design), but provided each new problem is approached in the way that has been described in this chapter the interface should give maximum possible efficiency.

6
Machine controlling man

There are many situations in industry in which it is the machine or conveyer which controls the rate at which an operative can work. This is known as paced work, which has been defined by Murrell (1965b) as 'any system when an operative is induced, either mechanically or by adjacent operatives, to work at a rate other than that of his/her own choosing'. As a general rule the mechanical inducement will occur when an operative is working on a single machine or at a work station on a moving belt; pacing will usually be operative-induced when a group of operatives are working on a production line which is not mechanically controlled and the rate of arrival of components at one operative is controlled by the rate of work of the preceding operative. The main difference between the two types of pacing is that when pacing is by a machine the sequence of events involving the operative will occur at regular intervals, whereas when a task is operative-paced the sequence will occur at intervals which are irregular and unpredictable.

Pacing has been investigated both in the laboratory and on the shop floor. The results are difficult to interpret and compare because there appears to be no unanimity in terminology and in some instances it is quite difficult to determine exactly what the experimenter demanded of his subjects. This led Murrell (1965b) to propose a classification of pacing which not only standardized the terminology but also identified six different types of pacing. He pointed out that there could be comparability only between the results of research which compared pacing of the same type. These six types of pacing were divided into two rigid systems, three semi-rigid systems and one non-rigid system. It is the two rigid systems which put the greatest constraint on the operative.

In Type 1, the operative has a fixed time (the 'dwell') in which to carry out an operation such as feeding. If he fails to complete this operation before the machine indexes an empty position results (a 'miss'). The operative works intermittently, having a 'wait' element in each operation cycle. This kind of pacing is produced by a type of machine found in many light engineering factories. It is basically circular in design and has a series of heads in each operating position. The transport device indexes at regular intervals, the length of the interval being determined by the time required to carry out the longest of the machining operations. An operative sits at one work station and her job is to unlock a jig, remove and dispose of a completed component and to reload and relock the jig; this she must do while the machine is stationary. If she completes her task in time all is well, but if she does not, according to the design of the machine either an empty jig will move round the machine or she is able to override the indexing by means of a pedal which will hold the machine until she completes her task. In either event, there is a loss of production.

In Type 2 machine-controlled pacing the operative working continuously has a task to carry out which must start and be completed during an interval of time (the 'tolerance') which is determined by the machine. If she is unable to complete the task before the end of the tolerance, she will 'miss' and have to complete at the beginning of the next tolerance. An example of this kind of machine is that used for cigar wrapping. The operative has to spread the wrapper leaf over a suction die in such a manner as to reduce the waste of this expensive component to a minimum. She must then get her hands out of the way of an arm which swings over to pick up the wrapper leaf and carry it into the machine where it is wrapped round the filler of the cigar. The time allowed for this operation is likely to be about 4 secs and there is considerable judgement involved in the careful and exact positioning of the leaf. If the task has not been completed in time, the operative can either override the machine by stopping it or she can jerk the leaf back out of the way so that the arm returns without a wrapper leaf attached to it. In either event, production is again reduced.

In these two types of pacing in particular, operative variability plays a very important part. How this happens is illustrated in Fig. 6.1. If the operative were allowed to do the job unpaced, she would produce a frequency distribution of service times about a mean ($\bar{X}$). (Service time is the time which an operative takes to execute a particular complete cycle of elements, excluding delays or waits, making up a task.) Now if the machine is running at the mean service time of the operative and if the tolerance allowed by the machine is quite small, this can be represented by two lines running parallel to the mean. If the service time falls within these two limits, then all will be well, but when it is longer than the upper limit the task will not be completed in the time allowed. A 'miss' will therefore occur. If, on the other hand, the service time is faster than the lower limit, no advantage accrues because the operative has to wait until the commencement of the next tolerance; so those 10 per cent of occasions on which the operative completes a task very quickly cannot compensate for those 14 per cent of occasions on which she completes the task slowly. In practice, too many misses are unacceptable and so the machine cycle time is likely to be set below the natural mean of the operative's service time so as to shift the upper limit of the tolerance closer to the tail of the distribution. In this example a shift of 0·6 sec would reduce the misses to 3 per cent. This clearly means that the effect of

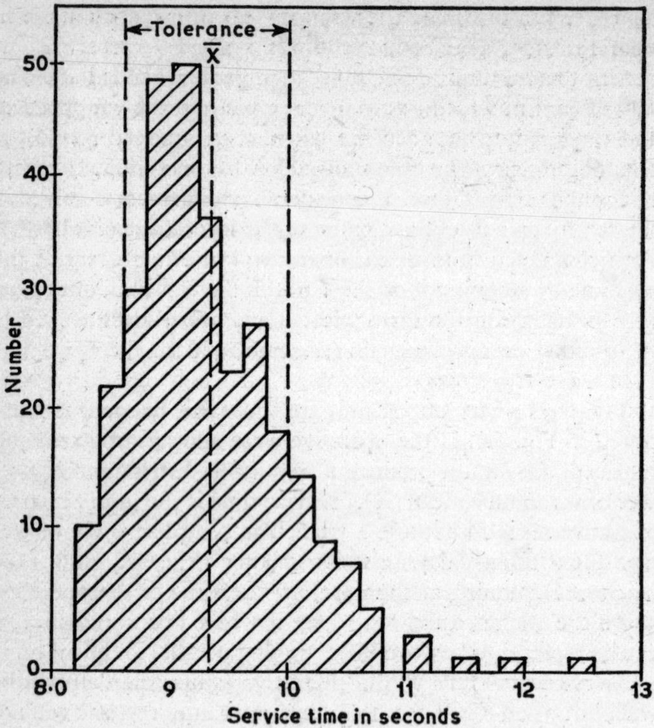

Fig. 6.1 *Histogram of service times showing the relationships of the mean to a tolerance*

pacing is to reduce the rate of production to 9 per cent below what it would be if the operative were free to go at her own pace. If the requirements of the process are such that it cannot be speeded up to a level approaching the operative's mean so that the machine cycle time falls in the tail of the distribution of the operative's service time, then there is no loss as a result of having a paced situation since the machine is the limiting factor on the rate of production. But if the machine is capable of going a good deal faster than the operative's mean and its rate has to be kept down to match that of the operative, then it is the operative who is acting as the limiting factor.

Since many machines have some form of 'hold' facility, whereas in others it does not matter if there are misses, it seems to be generally accepted by management that it is a good thing to run machines rather faster than they should be for a match to the operative. The idea behind this practice seems to be that if an operative is forced to work faster than she would do naturally her whole performance will inevitably speed up. However, in studies of machine pacing in the laboratory Murrell (1963) has demonstrated that this is not so; the forced pace does not produce a change in the mean or in the distribution of the service times.

This failure is reflected in the machine efficiency; if an operative misses on 20 per cent of occasions the machine efficiency will be 80 per cent. But this is a deceptive measure; a machine running at 200 units/hour with a machine efficiency of 95 per cent will produce 190 units whereas another running at 300 units/hour at 80 per cent efficiency will produce 240 units. So machine efficiency in the pacing situation is a measure of the mismatch of the man and machine and not a measure of productivity.

An extensive study of machine pacing was carried out under the auspices of the IHRB by Wyatt and Langdon (1938). They state that most of the firms they visited deliberately run their machines at a speed which is faster than that with which the operative can naturally cope. They give as an example a machine on which small steel chains were being manufactured which was run at 37 units per minute with an average machine efficiency of only 64·3 per cent; even during the best hour of the day it reached only 68·2 per cent. They give an output curve which shows a fairly steady fall of output throughout the working day, with the output for the last hour nearly 20 per cent below that of the highest level recorded. Wyatt and Langdon suggest that this kind of situation causes a good deal of strain and that the task is disliked by the operatives. They cite twenty-nine jobs which they studied, in all of which the machine speeds were running faster than that with which the operatives could cope and on none of which was the machine efficiency higher than 86 per cent, the lowest being 52 per cent.

Wyatt and Langdon's suggestion that excessive speed causes strain and distaste among operatives is reinforced by an incident in the early sixties. A large firm making a nationally known product became concerned at the high rate of labour turnover amongst the girls in the packing department. Since they prided themselves on being good employers (as they were), they called in outside help to investigate. After interviewing a large number of former employees, the investigator reported that the main reason for leaving the firm was the high speed at which the packing line was run. Discussions followed with work study engineers who insisted that their times were correct and that the packing belts could not be running too fast. In any event, they claimed, it did not matter since any unpacked material which the girls could not cope with could be recycled. The work study engineers were right within the limits of their knowledge: they had established a mean cycle time which was reduced to provide opportunity for earning bonus and, ignoring the fact of operative variability, they expected every unit to be packed in this time. The pressure which this put upon the girls, however, was such that, as they got nowhere by argument, they were unwilling to continue to work on the job.

Another complication is that different operatives will have different variabilities, often about a similar mean performance time. The effect of this will be that operatives who have a smaller variability are likely to have a smaller number of 'misses' than those who have the greater variability and, as a consequence, their output will be higher. This in turn may influence the setting of machine speeds, since in practice individual machine speeds will not normally be set to correspond with the individual variability of the various operatives; they are more likely to be set to correspond to the performance of the less variable operatives so that the more variable will be somewhat overloaded, or vice versa.

An industrial study on the effect of individual variability upon performance is reported by Conrad (1955). The task investigated was the operation of an automatic weighing machine, which allowed only a set time for the operative to insert material. If she had not completed the task by the end of the tolerance, she would have to wait for the next cycle to come round. Conrad gives an example of two operatives both of whom had a mean cycle time of about 6 sec. One of them had a coefficient variation of 0·15 while the other was 0·30. Experimentally, he ran the weighing machines at a range of speeds between 4 cycles per min to 12 cycles per min and measured the effect on the two operatives' performance. The outcome is shown in Table 6.1.

Table 6.1. Effect of distribution of operative cycle time on output (after Conrad, 1955)

Machine		*Operator A (C.V. = 0·15)*		*Operator B (C.V. = 0·30)*	
Cycle time (seconds)	*Rate per Minute*	*Output per Minute*	*Percentage of Maximum*	*Output per Minute*	*Percentage of maximum*
15	4	4·0	100	4·0	100
10	6	6·0	100	5·9	98
8	7½	7·5	100	6·6	87
6	10	8·6	86	7·5	75
5	12	6·8	56	7·8	65

Unfortunately, Conrad does not give the tolerance time so the complete interpretation of his results is not feasible, but it will be seen that operative A was able to maintain 100 per cent production by speeding up somewhat. On the other hand operative B, with a higher variability, could not maintain 100 per cent production even when the machine cycle time was the same as the mean of her service time.

If the table is looked at again, it will be seen that the output per minute for operative A takes a downward turn when the rate per minute is increased from 10 to 12. That for operative B does not take a downward turn although the rate of increase is slowing down. As will be seen below, this effect was found also by Wyatt and Langdon (1938).

There is a very good theoretical reason for this which is set out in Murrell (1963). It is well known that in practice operator

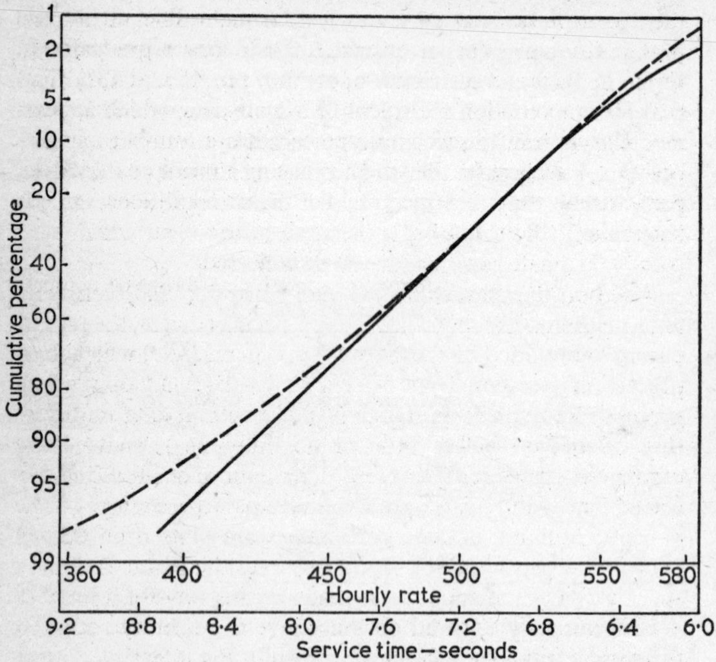

Fig. 6.2 *Cumulative frequency plotted against service time* (broken line) *and hourly rate* (solid line) (*from Murrell, 1963*)

variability is not even approximately normally distributed. Histograms of service times will be positively skewed (see A8, p. 48). Almost invariably the method of measurement is the duration of the service time and this is linearly related to the reciprocal of the rate per hour. By carrying out a harmonic transformation (i.e. converting to reciprocals) so that the cumulative frequency is now expressed in terms of rate per hour the skewedness, provided it is not too pronounced, is substantially reduced and the distribution is brought close to normality, especially at the lower rates of work which are of particular interest. As a consequence the cumulative frequency can be expressed as very nearly a straight line on normal probability paper (paper on which the Y axis is graduated in terms of a normal distribution, see A8, pp. 46 and 131) (Fig. 6.2). This method of treatment of output data, which appears not to have been previously reported, has a number of other practical uses besides the study of pacing since it enables data, particularly that obtained under field conditions, to be 'normalized' by the process of curve fitting even when comparatively small samples have been collected.

Based on the untransformed data Murrell (1963) constructed a mathematical model for Type 2 pacing (that is, the type of pacing exemplified by the cigar wrapping machine) which, by a process of iteration (that is, successive recalculation which successively reduces the range of the resulting data until further calculation makes little or no difference), enabled the number of misses at different rates per hour to be predicted. He found that if the misses are expressed as a percentage of the machine rate and if these percentages are plotted on normal probability paper against the hourly rate the result is again a straight line. Predictions were made in this way for a number of subjects which proved on the whole to be pretty accurate; the results from one subject are shown in Fig. 6.3.

Now it is a mathematical consequence that if data presented as percentages form a straight line on normal probability paper they will be converted into a curve with a maximum if they are expressed as actual output per hour against machine rate. This is shown in Fig. 6.4 for seven of the subjects tested by Murrell. Each curve shows a maximum and then begins to decline as the hourly machine rate is increased. With the exception of subjects B and D, the optimum machine rate to produce the greatest hourly output is different for all the subjects. Thus if the hourly machine rate had been set at 650 to suit subject G, the performance of subjects A to D would be hopelessly bad and well below the maximum of which they were capable. It seemed

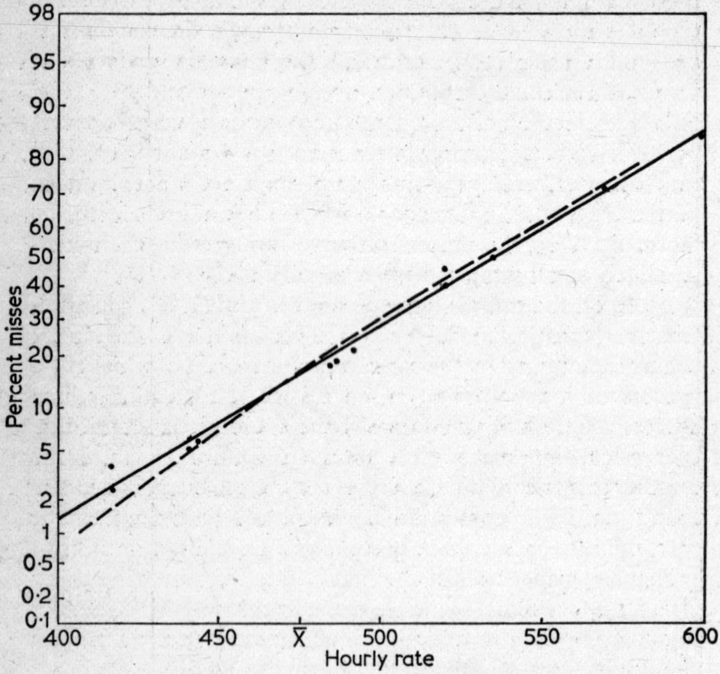

Fig. 6.3 *Predicted misses* (broken line) *and actual misses* (solid line) (*from Murrell, 1963*)

probable that subject A would be unemployable on this particular task at that speed. This emphasizes the importance, when tolerance ratio (that is, the ratio of the tolerance time to the machine cycle time) is small, of tailoring the machine speed to individual operatives, and shows that no one universal machine

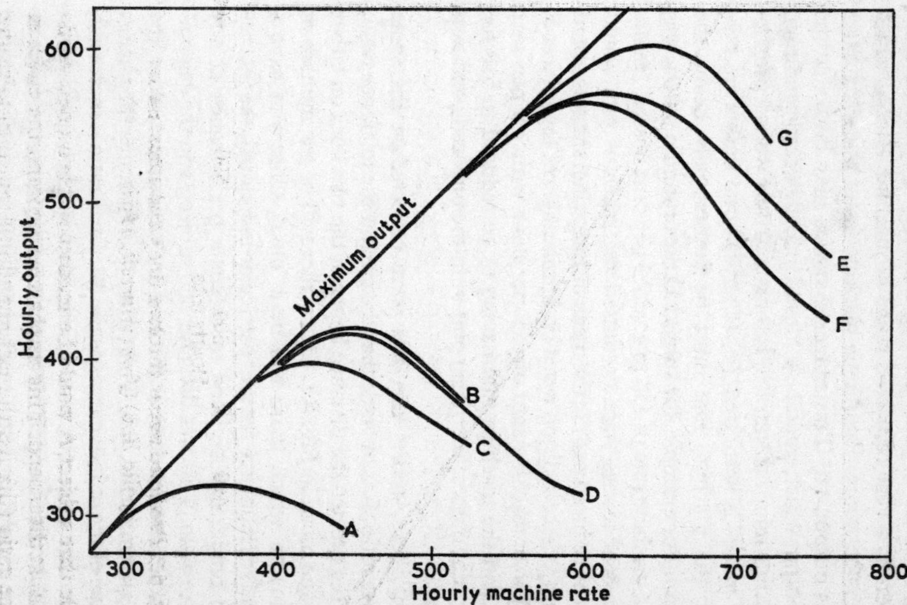

Fig. 6.4 *Seven subjects – the effect on output of changing the machine rate (from Murrell, 1963)*

speed would be economically viable if a group of industrial personnel showed the same differences as were found in Murrell's experiment.

The outcome of this investigation may be summarized: (a) There is a very restricted range of machine rates at which output is optimum. Running a machine faster than this range reduces output. (b) This range of speeds may be below the mean output which could be achieved in a self-paced condition. (c) The maximum output will be obtained only when a proportion of misses is permitted. (d) If virtually no misses are permitted, the machine rate must be substantially below the optimum. For example, for subject D the optimum output was 417 at a machine rate of 450 to give 7·3 per cent misses. If only ·05 per cent misses were permitted the machine rate would have to be reduced to 377; the mean output when she was tested in the unpaced condition was 475. Similar figures for subject E are 571 at a machine rate of 615 per hour with 7·8 per cent misses and the machine rate having to be reduced to 518 for ·05 per cent misses; her mean output in the unpaced condition was 620.

In all these studies the data were collected at the same tolerance ratio. It is obvious that the greater the tolerance ratio, that is the greater the distance between the two lines on either side of the mean in Fig. 6.1, the smaller will be the number of occasions on which there will be a 'miss'. Once you have a cumulative frequency for any particular individual, the mathematical model can be used to estimate performance at any tolerance ratio. This was done for one subject, and the calculations were subsequently confirmed by experiment, with the results shown in Fig. 6.5. This figure demonstrates the changing shape of the output curves as the tolerance ratio is increased, but even at a tolerance ratio of one, that is when one feeding position is just disappearing as the next appears, the effect of operator variability is still marked and it is not until tolerance ratios become greater than one that the full effect of operator variability is mitigated

Some explanation is required of the shape of the ·09 curve. This drops to a minimum and then starts to increase again as the machine rate increases. This is because a point is reached when the machine rate is so fast that the operative is completing her cycle of work only at every other tolerance, so effectively she will be working at half the indicated machine rate. When this happens, it would seem that a production curve will

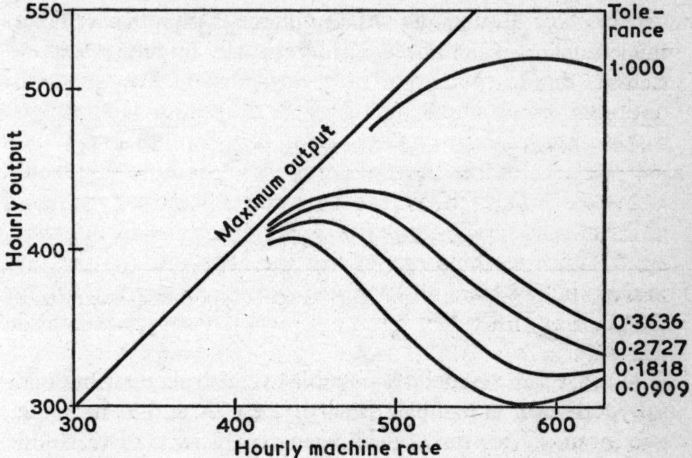

Fig. 6.5 *The effect on output of altering the tolerance (from Murrell, 1963)*

steadily increase up to its previous maximum and then decline again and so on; this has not actually been tested experimentally.

It will be clear that in Type 2 pacing the tolerance ratio is a critical variable and so in any experiment which compares performance at different rates of work the tolerance time must be maintained as a constant proportion of the machine cycle time, otherwise the experiment will be confounded. This seems to have happened in an experiment conducted by Buffa (1961), who used five machine rates with three 'times available'. Out of the fifteen possible tolerance ratios which were obtained from these variables, six are paired so that there are no less than twelve different tolerance ratios varying from ·13 to ·8. Since Buffa made no allowance for this variable (two degrees of freedom are used in the analysis of variance instead of eleven) the experiment is confounded, and his misleading finding that 'time available was not significant at any of the measurements made' is not surprising. In fact the experiment was a complete waste of effort. Even now this point does not seem to have been sufficiently appreciated by some investigators into the effects of machine pacing.

Subject	*Mean service time*	*Standard deviation*	*C.V.*
X	7·75 sec	1·89 sec	0·233
Y	7·75 sec	1·12 sec	0·144
Z	7·35 sec	1·06 sec	0·144

Table 6.2 Performance data for three hypothetical subjects (after Murrell, 1963)

Using the same model it is possible to construct distributions for hypothetical individuals. Two of these, X and Y, have the same mean service time but differing coefficients of variation (the coefficient of variation is the standard deviation divided by the mean; it enables the variation of data with different means to be compared), while Y and Z have the same coefficient of variation but differing mean service times, as is shown in Table 6.2. Their output curves are shown in Fig. 6.6. The shapes of the output curves of Y and Z are very similar, with Z giving a higher output than Y and at a higher machine rate. X, who had the same mean service time as Y but a greater variability, produces her maximum performance at about the same machine rate as Y but this maximum is substantially lower. These curves reinforce the point made above when quoting the findings of Conrad (1955) that it is the variability which appears to be an important determinant of paced performance; the reduction in the mean service time is also important but only if the standard deviation falls as well so that the coefficient of variation remains much the same.

So far we have been considering two types of rigid pacing and it is clear that an important determinant from the design point of view is that the tolerance ratio should be as large as

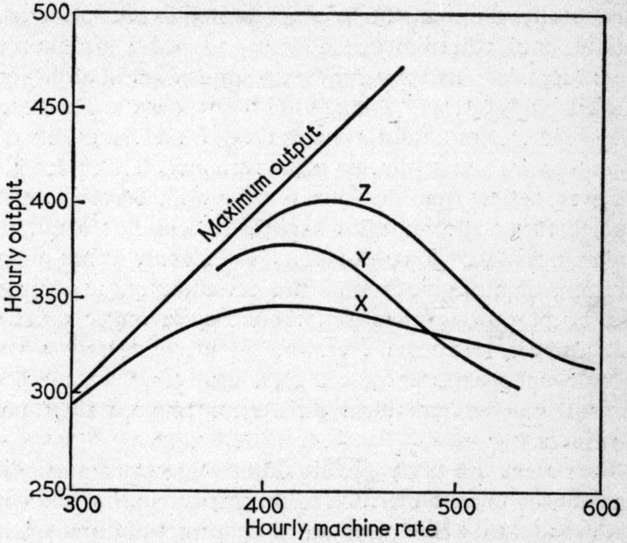

Fig. 6.6 *Effect on output of changing standard deviation and mean service time (from Murrell, 1963)*

possible; this can be achieved by increasing the number of work stations which are available to the operative at any one time. For instance, if an operative is feeding discs into holes on a moving belt the tolerance ratio can be increased by increasing the number of holes which are available to the operative at any one time, or in the case of the multi-head machine which has been mentioned several times previously, the jig could be available to the operative for unloading and loading over two intervals between indexing instead of only one. There are other ways in which the effects of rigid pacing may be reduced, thus converting the rigid systems into semi-rigid ones.

In the first of these the constraint is taken away either from the start or the end of the cycle (Type 3) or the components are allowed to accumulate in front of an operative in order to produce a buffer (Type 4B). In Type 5, which is found only occasionally, the components are presented to the operative at variable but predictable time intervals and she has to process all the components. As far as can be determined, most of the jobs studied by Wyatt and Langdon (1938) appear to have been semi-rigid systems, but the details they give do not enable this to be determined with any great certainty. This does not, however, detract from the value of their work although unfortunately they do not appear to have realized, as did Conrad, the importance of individual variability, so that only on one or two occasions do they give the standard deviations of the performance of the various operatives whom they investigated. They did, however, realize that even when groups of operatives have become highly experienced and apparently very efficient, there are still marked individual differences between their performances.

They seem also to have realized that there was a possibility that slowing down a machine might increase output and with this in mind they conducted an experiment with three operatives on a chain-making process which appears to have been Type 2. The subjects were three adult research workers who were given a practice period lasting for eighteen weeks, at the end of which their learning curves showed practically no further improvement. They were then given tests at machine speeds varying from 28 to 36 units per minute for work periods lasting for two and a half hours. The whole series was replicated once. The results showed that as the machine speeds increased so did the standard deviation of each subject, while their percentage efficiency decreased. From the data Murrell (1965a) correlated the standard deviation of each of the three subjects at the various speeds with the machine efficiency, using Kendall's tau; the resulting correlations are, for subject A, $\tau = -0{\cdot}94$, for subject B, $\tau = -0{\cdot}84$ and for subject C, $\tau = -1{\cdot}0$. These data show very neatly the very close relationship between the standard deviation and the output as a proportion of the machine speed. This output increased as the speed increased but at the higher speeds the output of two of the three subjects started to decline. Subject A reached his peak at 35 units per minute, subject B at 34 units per minute and subject C at 36 units per minute.

As a result of this Wyatt and Langdon seemed satisfied that a reduction in machine speed could improve output. They therefore carried out an experiment on the shop floor with the wrapping of chocolates. In this task it was essential that every space in the machine was filled; if the operative missed she had to override the machine and fill any empty slots before restarting. In the experiment the machine speed was progressively changed from 68·5 to 85·7 units per minute, a relative speed increase of 25 per cent. Two operatives were used and their relative increases in output at the highest speed tested were 2·6 and 7·2 per cent respectively. However, they achieved their greatest increase when the machine was running at 81 units per minute at which speed they showed 8 and 9·5 per cent increases respectively.

Analysis was also made of the time lost due to the operatives overriding the machines. Even at the slowest speeds some time was lost in this way and this showed a very sharp increase when the machine speed reached 78 units per minute in the case of one operator, with a more and gradual increase in the case of the other. There was a further substantial increase when the machine speed was advanced from 81 to 85·7 units per minute; one operative lost 10 per cent due to overriding while the other lost 6·1 per cent.

In another study of chocolate wrapping on machines capable of running at 240 units per minute, the machine speed was increased in steps of 15 units from 150 to 210 units per minute.

Four workers were tested for two days at each of the five machine speeds. As before, it was found that there was an optimum speed for each machine but that it was different for the different operatives. One operative produced her best output at 195 units per minute, two at 180 units per minute, and one at 165 units per minute. Unfortunately, the standard deviations of the various operatives are not indicated.

Had the investigations of Wyatt and Langdon and of Murrell been done in chronologically the reverse order, it could be claimed with some justification that Wyatt and Langdon confirmed in shop floor investigations what Murrell had already demonstrated in the laboratory. As it was, their work was done twenty-five years earlier and although the investigations of Murrell showed the same effects that Wyatt and Langdon had shown, Wyatt and Langdon did not have the benefit of this knowledge and worked primarily from intuition. That their intuition was right shows great credit to these particular investigators, even though they were working largely 'by the seat of their pants'.

The importance of operatives' variability has already been stressed and discussion has proceeded as if this remained constant throughout the working day. But this is almost certainly not the case, as Conrad and Hille (1953) pointed out when they reported on a study of the packing of glass jars. The working day was divided into four almost equal parts by breaks and they recorded the times of a number of work cycles about the middle of each work period. Five of the operatives were studied. Looking at the differences between the mean packing time in the first and fourth periods, we find that only in one instance was there any substantial change in the rate of work – one operative's mean packing time increased by about 11 per cent. In the other instances the differences are so small and of such variable direction that they are negligible. This is not true, however, of the coefficient of variation. One operative who was very consistent showed an increase of only 5 per cent, but the coefficient of variation of three others increased by over 20 per cent with the remaining one increasing by nearly 15 per cent. This was of course a self-paced task, otherwise the changes which were observed could not have been recorded.

Similar results had been obtained thirty years earlier by Wyatt and Fraser (1925) who carried out a study of the folding of handkerchiefs. The work lasted four and a quarter hours and from the data given in the report it is possible to calculate means and standard deviations for three of their subjects. These are shown in Table 6.3. The means and standard deviations remain reasonably stable for A and B up to 02·20 hours from the start while C, who was a slower, less experienced worker, begins her drift after 01·44 hours. But in the period ending at 03·05 all three have shown substantial changes, the standard deviations deteriorating more than the means as the sharp increases in the coefficients of variation show.

Table 6.3 Means, standard deviations and coefficients of variation of the operatives employed on handkerchief folding (from Murrell, 1962)

		Time from start of work			
		00·36	01·44	02·20	03·05
A	*Mean*	110·0	102·0	104·5	121·5
	SD	7·75	9·80	11·53	17·38
	CV	·070	·096	·110	·143
B	*Mean*	106·0	103·0	107·0	129·5
	SD	8·60	10·05	9·54	15·87
	CV	·081	·098	·089	·123
C	*Mean*	142·5	112·0	132·0	135·0
	SD	14·35	11·66	14·0	22·14
	CV	·101	·104	·106	·164

It is possible to argue that if variability will increase throughout a working period and output will depend on this variability, more satisfactory results will be obtained if the speed of machines is varied also. This line of thought occurred to Wyatt and Langdon, but for different reasons, and they carried out a further experiment on the chocolate wrapping machines. The operatives normally worked four hours at a spell and there was a decrease in output throughout the period. Two operatives took part in the experiment. In the first condition the machine

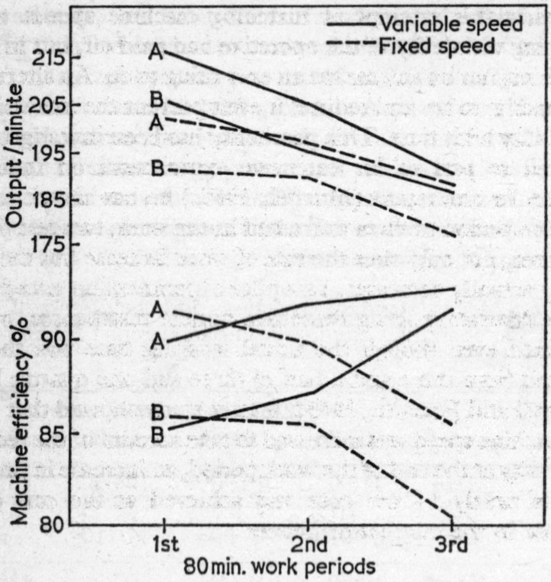

Fig. 6.7 *The effect on output of changing machine speed during a shift (from Murrell, 1963; based on data from Wyatt and Langdon, 1938)*

was maintained at a uniform speed of 220 per minute, which was the normal speed for this machine. In the other condition the machine was run for the first 80 mins at 240 per minute, for the second 80 mins at 220 per minute and for the third 80 mins at 200 per minute. The outcome is shown in Fig. 6.7. Output declined under both conditions but more when the speed was varied, but as it started from a higher level the total output of subject A was up by 4·1 per cent and that of subject B by 5·4 per cent. Machine efficiency, which is shown in the lower part of Fig. 6.7, gives the truest indication of what has occurred. When the variable speeds were used, machine efficiency increased throughout the working period instead of decreasing as it had done previously.

Clearly this attempt at matching machine speeds to the changing variability of the operative had paid off, but in practice this is not by any means an easy thing to do. An alternative approach is to try and reduce or even prevent the increasing in variability with time. This possibility has been investigated by Murrell as part of his extensive experiments on industrial fatigue. In one report (Murrell, 1965c) he has shown that if, during a period of three and a half hours work, two rest pauses are given, not only does the rate of work increase but the variability actually decreases. In a pilot experiment on paced work in the laboratory using three five-minute rest pauses, output increased even though the actual working time was thereby reduced from three and a half to three and one quarter hours (Murrell and Forsaith, 1963). A later study showed that when the machine speed was increased to take account of the reduced variability at the end of the work period, an increase in production of nearly 14 per cent was achieved at the cost of an increase in the number of 'misses'.

Production lines

Production lines vary very much in their method of operation. One quite common type has a series of work stations along a belt which may be moving intermittently or continuously, allowing each operative a fixed time to complete the allotted task. This arrangement is in effect a series of individual paced work stations strung together and, as a rule, the performance of one operative has no effect on the performance of his neighbours. The pacing, therefore, should be regarded as being one of the types which have already been described. If the pacing is very tight, it is possible for one operative to influence another in that it is impossible for an operative to override a conveyer or a belt if he cannot complete his cycle of operations in the space which has been allocated to him. Therefore, in order to complete his task, he would have to encroach upon the space of his next-door neighbour down the line, who in turn will be prevented from starting his task at the point at which it should become available to him. As a result of this you can get a kind of ripple effect down a production line which can push everyone out of station and this can usually only be restored by bringing in extra hands temporarily. In order to avoid this highly undesirable situation, the line will often be run sufficiently slowly to ensure that each operative will be able to complete the allotted task without encroaching upon his neighbour and this in turn means that production will be lower than it could be. One remedy for this is to allow sufficient space between work stations to act as a buffer to absorb the operative variability. The disadvantage of this arrangement is that the gains obtained by increasing the rate of production may be outweighed by the cost of the additional space and equipment needed.

This point was investigated by van Beek (1961) on a rather different type of production line (Type 5). In this type of pacing components are moved from one work station to the next by hand. When the components are quite small, it is possible to provide buffers between each work station the size of which can be calculated when the mean and standard deviation of the work content is known. It is often overlooked by work study engineers that the process of balancing a line like this does not consist only of ensuring, as far as possible, that the work content of each station is the same, but also of investigating the effect of the operator variability on the performance of the line. When the components are very large, it may be impracticable to provide buffers for the same reason as it would be on the conveyer assembly line just referred to, but under these circumstances, as also on the conveyer, it is possible with the use of a computer to investigate the probable pay-off for making various changes.

This is just what van Beek did on a television assembly line which consisted of 104 girls. There were no buffers between the girls so that girl A could not pass her completed chassis to girl B until girl B had already passed hers to girl C and so on. The first thing van Beek did was to investigate the variability of the individual operatives and then, using a computer, he calculated the amount of time lost through girls simply having to wait either for a set to be passed to them because the girl preceding her was having one of her long cycles or through having to wait to pass her completed job onto the girl following her who couldn't receive it, either because the girl beyond her again was not ready or because she herself was taking longer than average. Losses as a result of this were estimated at over 17 per cent and van Beek next estimated what the effect would be if he were to put one buffer set between every other operative and break the team down into four units of twenty-six with a somewhat larger buffer stock between each unit. The losses came down to about 4 per cent; the gain resulting from putting one set between each girl was an improvement of only just over another 1 per cent. The savings which would result from substantial reduction in lost time were estimated and were set against the additional cost of space and equipment which would be required to carry out the changes which were investigated. Van Beek found that the break-even point was when there was one set between every other girl; the extra cost of putting a set between each girl was more than the savings which would result.

Van Beek is a psychologist working for the Philips Organization in Eindhoven and his report illustrates very nicely the kind of investigation which can be carried out by someone with psychological training in an industrial environment. Actually, although it is not strictly relevant to the subject of this book, he also investigated the social changes which were the result of breaking the large team of 104 down to four units of twenty-six each. The smaller units were very much preferred by the girls involved.

The simulation of production lines in this way is comparatively simple though it does involve measuring for preference or assuming, if this is impossible, the standard deviation around the mean of the work content of each station; and the same technique can be used to determine the optimum buffer size for operatives working on light, small components. Unfortunately, this kind of modelling only seems very rarely to be used.

In this chapter we have looked at the situation where, in effect, the machine is controlling the man, one being fixed and the other being variable. The success with which this is done depends on there being a clear understanding of all the factors involved and so designing the work situation that the operative can meet the requirements of the task without undue stress. When more than one work station is involved, this can best be done by simulation of the situation on a computer.

Summary

We have seen that for more than sixty years psychologists have been involved in the study of human work; one aspect of which has been discussed in some detail in this book, while another – how can employees be motivated to work well? – is reviewed in E5. The period has been one of accelerating technological change which has produced increasing demand on the mental rather than the physical capacities of operatives, this in turn has thrown up a range of problems which can be resolved only by use of psychological knowledge. But having this knowledge is not of itself sufficient – it has to be turned into practical reality and this involves engineers, work psychologists, work study practitioners, personnel managers and others. This interdisciplinary attack on practical work problems was in evidence in a small way before 1939 and became more formalized in 1949 with the formation of the Ergonomics Research Society.

There is no doubt that research into man's capacity to operate machines has resulted in design recommendations which have increased efficiency and greatly reduced some of the more blatant abuses of man by machine. But even more important, it is beginning to develop a different outlook and understanding amongst designers, administrators and even legislators of human capacities and limitations. With increasing frequency, promotional material is claiming that a particular piece of equipment is designed in accordance with ergonomic, human factors or (more rarely) psychological principles. There is still much to be done and the role of achievement will depend upon the extent to which others involved in work design accept psychologists as technologists with a pool of useful and practical knowledge, rather than as 'head shrinkers' or 'rat chasers'.

References and Name Index

The numbers in italics following each entry refer to page numbers within this book.

Baker, C. H. (1962) *Man and Radar Displays*. Oxford: Pergamon Press. *67*

Barbour, J. L. and Garner, W. R. (1951) The effect of scale numbering on scale reading accuracy and speed. *J. Exp. Psychol. 41*: 298. *55*

Bartlett, F. *11*

Birmingham, H. E. and Taylor, F. V. (1954) A design philosophy for man/machine systems. *Proc. Inst. Radio Engineers 42*: 1748. *26, 27, 31*

Bornemann, E. (1942) Untersuchungen über der grad der geistigen beanspruchung. *Arbeitsphysiol. 42*: I Aus arbeitung der methode *142*: II Praktischergebnisse 173. *27*

Broadbent, D. E. (1956) Listening between and during practical auditory distraction. *Brit. J. Psychol. 47*: 51. *28*

Buffa, E. S. (1961) Pacing effects in production lines. *J. Indust. Eng. 12*: 383. *121*

Burke, D. and Gibbs, C. B. (1965) A comparison of free-moving and pressure levers in a positional control system. *Ergonomics 8*: 23. *80*

Chapanis, A., Garner, W. R. and Morgan, C. T. (1949) *Applied Experimental Psychology*. New York: Wiley. *37, 44*

Christensen, J. M. (1952) *Quantitative Instrument Reading as a*

Function of Dial Design, Exposure Time, Preparatory Fixation and Practice. USAF Aero-Medical Lab. Rept No. AF52/116. *39*

Churchill, A. V. (1956) The effect of scale interval length and pointer clearance on speed and accuracy of interpolation. *J. App. Psychol. 40*: 358. *58*

Churchill, A. V. (1959) Optimal interval length for visual interpolation. The effect of viewing distances. *J. App. Psychol. 43*: 125. *49, 53*

Conrad, R. (1955) *Setting the Pace*. MRC App. Psychol. Unit. Rept No 232/55, Cambridge. *115, 122*

Conrad, R. and Hille, B. A. (1953) A comparison of paced and unpaced performance at a packing task. *Occup. Psychol. 29*: 15. *126*

Cook, T. W. and Shephard, A. H. (1958) Performance on several control display relationships as a function of age and sex. *Percept. and Motor Skills 8*: 339. *85*

Craik, K. W. (1947a) Theory of the human operator in control systems. 1: The operator as an engineering system. *Brit. J. Psychol. 38*: 56. *21*

Craik, K. W. (1947b) Theory of the human operator in control systems. 2: Man as an element in a control system. *Brit. J. Psychol. 38*: 142. *21*

Davis, D. R. (1958) Human errors and transport accidents. *Ergonomics 2*: 24. *104*

Elkin, E. H. (1959) *Effect of scale shape, exposure time and display complexity on scale reading efficiency*. USAF, WADC Tech. Rept 58–472. *42*

Fitts, P. M. (1947) Psychological research on equipment design in the AAF. *Amer. Psychologist 2*: 93. *12*

Fitts, P. M. (1951) Engineering psychology and equipment design. In S. S. Stephens (ed.) *Handbook of Experimental Psychology*. New York: Wiley. *21*

Fitts, P. M. and Jones, R. E. (1947) *Psychological aspects of instrument display. 1. analysis of 270 'pilot error' experience in reading and interpreting aircraft instruments*. USAF, Air Material Command, Wright-Patterson AF Base, Memo Rept No. TSEAA-694-12A. *90*

Garvey, W. D. and Taylor, F. V. (1959) Interactions among operator variables, system dynamics and task induced stress. *J. App. Psychol. 43*: 79. *28*

Gibbs, C. B. (1954) The continuous regulations of skilled response by kinaesthetic feedback. *Brit. J. Psychol. 45*: 24. *76, 80*

Gilbreth, F. *13*

Graham, Nora, E. (1956) The speed and accuracy of reading horizontal, vertical and circular scales. *J. App. Psychol. 40*: 228. *41*

Grether, W. F. (1947) *The Effect of Variations in Indicator Design upon the Speed and Accuracy of Altitude Readings.* Aero-medical Laby. USAF Air Material Cmnd, Rept No. TSEAA-694-14 (also in *J. App. Psychol. 33*: 363. 1949). *70*

Grether, W. F. (1948) Design of instrument dials for ease of reading. *SAE Quart. Trans. 2*: 539. *58*

Grether, W. F. and Connell, S. C. (1948) *Psychological Factors in Check Reading Signals.* USAF Air Materiel Command, Rept MCREXD 694–17A. *61*

Grether, W. F. and Williams, A. C. (1947) Speed and accuracy of dial reading as a function of dial diameter and angular separation of scale divisions. In P. M. Fitts (ed.) *Psychol. Res. in Equipment Design.* AAF Aviation Psychol., Prog. Res. Rept No. 9. *45, 47, 53*

Harper, R. (1952) Psychological and psycho-physical studies of craftsmanship in dairying. *Brit. J. Psychol., Monogr. Suppl. No. 28.* *24*

Herzberg, F. *13*

Hick, W. E. and Bates, J. A. (1950) *The Human Operator of Control Mechanisms. The permanent record of research and development.* Monograph No. 17.204. London: HMSO. *21*

Holmes, J. G. (1941) The recognition of coloured light signals. *Trans. Illum. Eng. Soc. VI*: 71. *64*

International Commission on Illumination (1959) *Colours of Light Signals.* Pub. CIE, No. 2 (W-1.3.3). Paris: International Commission on Illumination. *64*

James, D. G. and Murrell, K. F. H. (1967) The design of scales for test instruments. *Ergonomics 10*: 707. *60*

Jenkins, W. O. (1947) The discrimination and reproduction of motor adjustments with various types of aircraft controls. *Am. J. Psychol. 60*: 379. *81*

Jones, J. C., Ward, A. J. and Haywood, P. M. (1965) Reading dials at short distances. *AEI Engineering.* January/February: 28. *49*

Jones, R. E., Milton, J. L. and Fitts, P. M. (1949) *Eye Fixations of Aircraft Pilots IV. Frequency, Duration, Sequence of Fixations during Routine Instrument Flight*. USAF, AF Tech. Rept 5975. *106*

Kalsbeek, J. W. H. (1964) On the measurement of deterioration in performance caused by distraction stress. *Ergonomics 7*: 187. *28*

Kalsbeek, J. W. H. (1971) Standards of acceptable load in ATC task. *Ergonomics 14*: 641. *28*

Kappauf, W. E. and Smith, W. M. (1948) *Design of Instrument Dials for Maximum Legibility. II: A Preliminary Experiment on Dial Size and Graduation*. USAF Air Materiel Cmnd, Wright-Patterson AFB, Aero Med. Lab. Memo. Rept No. MCREXD-694-1N. *45, 47, 53*

Kappauf, W. E., Smith, W. M. and Bray, C. W. (1947) *Design of Instrument Dials for Maximum Legibility*. USAF Aero-Medical Lab. Memo. Rept No. TSEAA-694. *41, 43, 46*

Kelly, C. R. (1968) *Manual and Automatic Control*. New York: Wiley. *110*

Kuntz, T. E. and Sleight, T. B. (1950) Legibility of numerals: the optimum ratio of height to width of strokes. *American J. Psychol. 63*: 567. *59*

Kurke, M. I. (1956) Evaluation of a display incorporating quantitative and check reading characteristics. *J. App. Psychol. 40*: 233. *61*

Leyzorek, M. (1949) Accuracy of visual interpolation between circular scale markers as a function of the separation between markers. *J. Exp. Psychol. 39*: 270. *45*

Loucks, R. B. (1944) *Legibility of Aircraft Instrument Dials: A Further Investigation of the Relative Legibility of Tachometer Dials*. AAF School of Aviation Medicine, Randolph Field, Texas. Project 265. Rept 1. *43*

Loucks, R. B. (1945) *Evaluation of Aircraft Altitude Indicators on the Basis of Link Instrument Ground Trainer Performance*. USAF Sch. Aviat. Med. Randolph Field. Proj. 341. Rept 1. *91*

Loveless, N. E. (1956) Display-control relationships on circular and linear scales. *Brit. J. Psychol. 47*: 271. *94*

Loveless, N. E. (1962) Direction of motion stereotypes: A review. *Ergonomics 5*: 357. *93*

Lythgoe, R. J. (1932) *The Measurement of Visual Acuity*. MRC Special Rept Series No. 173. London: HMSO. *59*
McCormick, E. J. (1964) *Human Factors Engineering*. New York: McGraw-Hill. *79*
Mackworth, N. H. (1950) Researches on the Measurement of Human Performance *MRC Special Report Series No. 298*. London: HMSO. *68*
Meyers C. S. *11*
Morgan, C. T., Cook, J. S., Chapanis, A. and Lund, M. W. (1963) *Human Engineering Guide and Equipment Design*. New York: McGraw-Hill. *49*
Morley, N. J. and Suffield, N. G. (1951) *The Effect of Pointer Position on the Speed and Accuracy of Check Reading Groups of Dials*. Admiralty. Naval Motion Study Unit Rept No. 46. *61*
Münsterberg, H. (1913) *Psychology and Industrial Efficiency*. Boston: Houghton Mifflin. *11*
Murrell, K. F. H. (1951) *The Design of Dials and Indicators*. Admiralty. Naval Motion Study Unit Rept No. 43. *37*
Murrell, K. F. H. (1952) The use and arrangement of dials. *Instrument Practice 6*: 520. *38*
Murrell, K. F. H. (1960a). *Ergonomics – Fitting the Job to the Worker*. London: Brit. Productivity Council. *74*
Murrell, K. F. H. (1960b) A comparison of three dial shapes for check reading instrument panels. *Ergonomics 3*: 231. *61, 106*
Murrell, K. F. H. (1962) Operator variability and its industrial consequences. *Int. J. Prod. Res. 1*(3): 39. *127*
Murrell, K. F. H. (1963) Laboratory studies of repetitive work I. Paced work and its relationship to unpaced work. *Int. J. Prod. Res. 2*: 169. *113, 116 ff.*
Murrell, K. F. H. (1965a) *Ergonomics: Man in his Working Environment*. London: Chapman & Hall. *14, 21, 29 ff., 36, 37, 38, 40, 44, 45, 56, 62, 78, 86, 87, 95, 124*
Murrell, K. F. H. (1965b) A classification of pacing. *Int. J. Prod. Res. 4*: 69. *109 ff.*
Murrell, K. F. H. (1965c) Le concept de fatigue une réalité ou une gêne? *Bulletin du CERP 14*: 103. *129*
Murrell, K. F. H. (1971) Industrial work rhythms. In W. P. Colquhoun (ed.) *Biological Rhythms and Human Performance*. London: Academic Press. *28*
Murrell, K. F. H. (1973) Applied psychology in the production of a British Standard. *App. Ergonomics 4*: 29. *51, 57*

Murrell, K. F. H. and Forsaith, Bel (1963) Laboratory studies of repetitive work. II: Results from two subjects. *Int. J. Prod. Res. 2*: 247. *129*

Murrell, K. F. H. and Kingston, P. M. (1966) Experimental comparison of scalar and digital micrometers. *Ergonomics 9*: 39. *69*

Murrell, K. F. H. and McCarthy, C. (1951) *Direction of Motion Relationships between Valves and Gauges when Mounted in Panels – Results from a Printed Test*. Admiralty. Naval Motion Study Unit, Dept. No. 51. *98*

Murrell, K. F. H., Laurie, W. D. and McCarthy, C. (1958) The relationship between dial size, reading distance and reading accuracy. *Ergonomics 1*: 182. *47 ff., 57*

Naylor, G. F. K. (1954) An approach to the study of dial reading. *Occup. Psychol. 28*: 90. *41, 42, 58, 59*

North, J. D. (1956) The application of communication theory to the human operator. In C. Cherry (ed.) *Information Theory: Third London Symposium. 1955*. London: Methuen. *21*

Papaloïzos, A. (1961) Some characteristics of instrument measuring dials. *Ergonomics 4*: 169. *59*

Poulton, E. C. (1958) Measuring the order of difficulty of visual motor tasks. *Ergonomics 1*: 234. *28*

Poulton, E. C. and Brown, I. D. (1961) Measuring the spare mental capacity of car drivers. *Ergonomics 4*: 35. *28*

Rockway, M. R. (1954) *The Effect of Variations in Control–Display Ratio and Exponential Time Delay on Tracking Performance*. USAF, WADC, Tech. Rept, 54–618. *82*

Rolfe, J. M. (1969) Human factors and the display of height information. *App. Ergonomics 1*: 16. *70*

Schouten, J. F., Kelsbeck, J. W. H. and Leopold, F. F. (1962) Evaluation of perceptual and mental load. *Ergonomics 5*: 251. *27*

Schum, D. A., Robertson, J. R. and Matheny, W. G. (1963) *Altimeter Display and Hardware Development 1903–1960 – Interim Report*. Fort Worth: Bell Helicopter Co. Report No. ASD-TDR-63-288. *70*

Shackel, B. (1960) Electro-oculography: The electro recording of eye position. *Proc. Third Int. Conf. Med. Electronics* 323. *106*

Simon, E. W. (1951) Effect of stress on performance with

preferred and non-preferred instrument design. *Amer. Psychol.* *6*: 387. *88*
Sleight, R. B. (1948) The effect of instrument dial shape on legibility. *J. App. Psychol. 32*: 170. *39 ff.*
Soar, R. S. (1955) Height, width, proportion and stroke width in numeral visibility. *J. App. Psychol. 39*: 43. *57*
Soar, R. S. (1958) Numeral form as a variable in numeral visibility. *J. App. Psychol. 42*: 158. *57*
Spencer, J. (1961) Observation accuracy in relation to the BSI draft recommendation for industrial instrument scale design. *Brit. J. App. Physics 12*: 712. *50, 51*
Spencer, J. (1963) Pointers for general purpose indicators. *Ergonomics 6*: 35. *58, 59*
Suddon, F. H. and Link, J. D. (1959) Handedness, body orientation and performance on a complex motor task. *Percept. and Motor Skills 9*: 165. *96*
Taylor, F. V. (1947) Psychology at the Naval Research Laboratory. *Amer. Psychologist 2*: 87. *12*
Taylor, F. V. (1957) Simplifying the controller's task through display quickening. *Occup. Psychol. 31*: 120. *31*
Taylor, F. V. (1959) The human as an engineering component. *AMA Arch. Industr. Health 19*: 278. *21*
Taylor, F. V. and Garvey, W. D. (1959) The limitations of a procrustean approach to the optimisation of men/machine systems. *Ergonomics 2*: 187. *21, 88*
Taylor, F. W. *13*
Thomas, D. R. (1957) Exposure time as a variable in dial reading experiments. *J. App. Psychol. 41*: 150. *41*
Tustin, A. (1957) The nature of the operator's response in manual control and its implications for controller design. *J. Inst. Elect. Eng. (Lond.) 94*: 190. *21*
Van Beck, H. G. (1961) *Working on Assembly Lines*. Eindhoven: Philips. *130*
Van Cott, H. P. and Kinkade, R. G. (eds) (1972) *Human Engineering Guide to Equipment Design* (Revised edn). Washington D.C.: U.S. Govt Printing Office. *14*
Vernon, M. D. (1946) *Scale and Dial Reading*. MRC Applied Psychology Research Unit. Rept No. 59, Cambridge. *55, 58*
Wallis, D. and Samuel, J. A. (1961) Some experimental studies of radar operating. *Ergonomics 4*: 155–68. *68*
Warwick, M. J. (1947a) Direction of movement in the use of

control knobs to position visual indicators. In P. M. Fitts (ed.) *Psychol. Research on Equipment Design*, 137–46. AAF Aviation Psychol. Programme Research Rept No. 19. *91 ff.*

Warwick, M. J. (1947b) Direction of motion preferences in positioning indicators by means of control knobs. *Amer. Psychol. 2*: 345. *97*

Weiss, B. (1953) *Building 'Feel' into Controls: the Effect on Motor Performance of Different Kinds and Amounts of Feedback.* ONR Special Devices Centre, Tech. Rept SDC 241-6-11. *79*

Welford, A. T. (1952) The 'psychological refractory period' and the timing of high-speed performance – a review and a theory. *Brit. J. Psychol. 43*: 2. *26*

Welford, A. T. (1959) Evidence of a single-channel decision mechanism limiting performance in a serial reaction task. *Quart. J. Exp. Psychol. 11*: 193. *26*

White, W. J., Warwick, M. J. and Grether, W. F. (1953) Instrument reading III. Check reading of instrument groups. *J. App. Psychol. 37*: 302. *58*

Wyatt, S. and Fraser, J. A. (1925) *Studies of Repetitive Work with Special Reference to Rest Pauses.* IFRB Rept No. 52. London: HMSO. *127*

Wyatt, S. and Langdon, J. N. (1938) *The Machine and the Worker. A Study of Machine Feeding Processes.* IHRB Rept No. 82. London: HMSO. *113, 116, 124, 128*

Yllö, A. (1962) The bio technology of card punching. *Ergonomics 5*: 75. *102*

Subject Index